T0227847

Taniguchi Symposia on Brain Sciences No.10

STRATEGY AND PROSPECTS IN NEUROSCIENCE

Taniguchi Symposia on Brain Sciences No. 10
PROGRAM COMMITTEE

Osamu Hayaishi (Chairman), Masao Ito, Yasuji Katsuki,
Yutaka Sano, Yasuzo Tsukada, Hiroshi Yoshida,
Teruo Nakajima

The Taniguchi Foundation, Division of Brain Sciences
ORGANIZING COMMITTEE

Osamu Hayaishi (Chairman)
    Osaka Medical College, Takatsuki, Osaka 569, Japan
Masao Ito
    Department of Physiology, Faculty of Medicine, University of Tokyo,
    Tokyo 113, Japan
Yasuji Katsuki
    National Institute for Physiological Sciences, Okazaki, Aichi 444,
    Japan
Yutaka Sano
    Department of Anatomy, Kyoto Prefectural University of Medicine,
    Kyoto 602, Japan
Yasuzo Tsukada
    Department of Physiology, School of Medicine, Keio University,
    Tokyo 160, Japan
Hiroshi Yoshida
    Department of Pharmacology, Osaka University Medical School,
    Osaka 539, Japan
Teruo Nakajima (Secretary General)
    Department of Psychiatry, Kyoto Prefectural University of Medicine,
    Kyoto 602, Japan

Taniguchi Symposia on Brain Sciences No.10

STRATEGY AND PROSPECTS IN NEUROSCIENCE

Edited by
Osamu Hayaishi

CRC Press
Taylor & Francis Group
Boca Raton London New York

CRC Press is an imprint of the
Taylor & Francis Group, an **informa** business

Supported in part by the Ministry of Education, Science and Culture under Grant-in-Aid for Publication of Scientific Research Result.

Published jointly by
Japan Scientific Societies Press Tokyo
ISBN 4-7622-0533-8
              and
VNU Science Press BV Utrecht, The Netherlands
ISBN 90-6764-111-1

Distributed in all areas outside Japan and Asia between Pakistan and Korea by VNU Science Press BV Utrecht, The Netherlands

PREFACE

This volume is based on papers presented at the 10th International Symposium, Division of Brain Sciences, the Taniguchi Foundation, which was held in Kyoto in November 1986. It was an exciting conference that clearly showed the advances made in neuroscience in recent years and the prospects for the future.

This series of symposia was initiated about 10 years ago through the good offices of Mr. Toyosaburo Taniguchi, the President of the Taniguchi Foundation. This meeting has been known worldwide by its unique features; namely, the age limit of all the speakers are, in principle, below 40 years and the number of participants is limited to no more than 20. This principle is based on the original idea of Mr. Taniguchi himself to encourage and to foster relatively young neuroscientists and to ensure uninhibited exchange of discussions and personal contacts among participants.

However, in commemoration of the 10th anniversary of this series of symposia, we chose the theme "strategy and prospects in neuroscience" for 1986 symposium and asked Mr. Taniguchi, who kindly consented, to make an exception to the previous rule and to invite senior scientists as well as relatively young participants to discuss freely and give their overview on this interesting subject. The format of this meeting was, therefore, extremely flexible and informal. We welcomed plenty of uninhibited speculations, imaginations, and discussions to make this meeting a truely fruitful and lively one. The organizers of this conference hope that this volume will serve as a milestone in this fascinating and rapidly developing field of bioscience and will stimulate further experiments that will lead to new frontiers in neuroscience. I take this opportunity to express deep gratitude to all participants, the members of the organizing committee, and the Taniguchi Foundation.

Osamu Hayaishi

EDITORIAL NOTE

The Taniguchi Foundation's Division of Brain Science organizes international symposia and invites young, active domestic and foreign scientists with the objectives of fostering both the researchers and the development of neurosciences. Subjects of previous symposia have been: 1) Neurobiology of Chemical Transmission, 2) Neurobiological Basis of Learning and Memory, 3) Brain Mechanisms of Sensation, 4) Neurotransmitter Receptors: Biochemical Aspects and Physiological Significance, 5) Structure and Function of Peptidergic and Aminergic Neurons, 6) Neuronal Growth and Plasticity, 7) Transmembrane Signaling and Sensation, 8) Humoral Control of Sleep and Its Evolution, and 9) Molecular Genetics in Developmental Neurobiology. The symposium's 10th anniversary focused on prospects for the future development of neurosciences and was entitled "Strategy and Prospects in Neuroscience". Top scientists in the discipline were invited.

In the first of the five sections of this volume the plenary lecture of Dr. John C. Eccles on the overall topic is reproduced. The second section deals with molecular neurobiology based on neurochemical approaches. Neural development and differentiation mainly using morphological methodology comprises the third section, while the fourth focuses on neural plasticity studied by neurophysiological methods. The fifth section offers discussions on prospects of neuroscience from the respective stances of the participants.

We hope this publication will contribute to the future development of the field of neuroscience by providing a helpful guide for current and future research undertaken by young scientists.

Editorial Board

CONTENTS

Neural Growth

Neural Plasticity

Opinions on Strategy and Prospects in Neuroscience

PERSPECTIVES IN NEUROSCIENCE

1

STRATEGY AND SOME PROSPECTS IN NEUROSCIENCE

JOHN C. ECCLES

Max-Planck-Institut für biophysikalische Chemie, Göttingen, F.R.G.

I. STRATEGY

When our mind is not flooded by sensory experiences, a frequent happening is that some image arises in our imagination. This image may be evocative of other images and these of still more, as we experience the mundane process of daydreaming. However, if we have a richly endowed mind, these images may be of beauty and subtlety, blending in harmony. If we express this imagery in some language, verbal, musical or pictorial, we may have artistic or scientific creation. Such creative imagination provides to others insight and understanding, and it is one of the most profound of human activities. Monod states that imagination in the game of scientific research is very different from that in children's games in that it is played with extremely austere and demanding rules. It is a very specific kind of creativity which cannot be done without a very clear concept of a certain number of basic postulates that have to be respected, otherwise you are not participating in scientific creation.

Let us follow Popper in stating that every scientific discovery develops out of a problem situation that may be recognized as arising either out of an existing scientific belief or out of a new insight. We suffer from a dissatisfaction with the existing problem situation. So, a particular problem is identified and studied and may offer to a scientist a great challenge, because it arises in a field that he understands well. As I well know, such a problem can literally take you over. As Popper maintains, you should care for your problem or have a sympathetic intuition about it. You enter into your problem so as almost to become part of it. This may be a very enjoyable adventure, but on the other hand you may become haunted by it even as you sleep!

Normally one recognizes a scientific problem and studies all the relevant literature so as to know the full context of the problem and of any attempted solutions. Often one practises what is called *subjective simulation*, identifying oneself with the object of the problem, the electron, the ion, the atom, the molecule, the organelle, the synaptic spine, the cell, *etc.*. This gives the vivid imaginative insight that leads to ideas, which are subjected to critical evaluation by the process of scientific rationality. There may be many sequences of creative ideas rejected in thought experiments. It is the fun of the scientific game. In this one learns more and more about the problem, this expertise enabling one to criticize more effectively further tentative ideas of oneself and of others. There seems to be luck in eventually arriving at a creative idea that stands up to scientific criticism and that leads to testable predictions. There is a wonderful word, "serendipity" for this good fortune that comes to one unexpectedly. It was coined by Horace Walpole in 1754 on the basis of a fairy tale entitled "The Three Princes of Serendip" (the ancient name for Sri Lanka). Walpole wrote "As their Highnesses travelled they were always making discoveries by accident or sagacity of things they were not in quest of."

If one is hoping for a life of scientific creativity it is of the greatest importance to choose a field which is full of important scientific problems and to work in a scientific institute where there is a good atmosphere as well as good equipment. No amount of book learning on the role of creative imagination and scientific rationality can compensate for the lack of experience of working in association with a creative scientist and other associates. The intimate contact during the experiments, watching the results and relating them to the problem situation and to previous findings is of inestimable value. I had the wonderful advantage of working with the great neuroscientist, Sir Charles Sherrington, during my early formative years at Oxford, and I have tried to give a similar environment to the many research associates I have been fortunate to have, there being actually 14 from Japan.

In the "Art of the Soluble" Medawar writes: "No scientist is admired for failing in the attempt to solve problems that lie beyond his competence. The most he can hope for is the kindly contempt earned by the Utopian politician. If politics is the art of the possible, research is surely the art of the soluble." I agree with Popper's criticism that this is too narrow a definition. It is restricting science to what Kuhn calls "normal science." In contrast, scientists should dare to choose problems that may not be soluble or whose eventual solubility is very doubtful. Similarly, a mountaineer may be dedicated to the climbing of a mountain which is regarded as unscalable. He will make many discoveries in reconnoitering around its base scaling lesser mountains and learning more about why the mountain is beyond his competence, or, optimistically, how to plan the ultimate climb.

An important question is: how does one get good ideas in science? Popper suggests that you take all the ideas that come into your mind, eliminating those that don't stand up to criticism. It is a kind of natural selection, as in biological evolution. A good idea that survives is quite rare. Monod has told a true story of Einstein. A Parisian hostess achieved a great success in bringing together in her salon Einstein and the poet Paul Valery. Valery had the task of engaging Einstein in conversation about his creativity. So Valery started, "How do you work, and could you tell us something of this?" Einstein was very vague about it. He said: "Well, I don't know . . . I go out in the morning and take a walk." "Oh," said Valery, "Interesting, and of course you have a notebook and whenever you have an idea you write it out in your notebook." "Oh," said Einstein, "No, I don't." "Indeed, you don't?" "Well, you know, an idea is so rare." Yet on another occasion Einstein is supposed to have claimed that he had an idea every two minutes, but that they were almost all bad!

My own experience is that when I am searching for a good new idea, I fill up my mind with knowledge on the problem and my critical evaluation of the attempted solutions of that problem. Then I wait the outcome of the mental tension so created. Maybe I take a walk as

Einstein did or I listen to music. This procedure is called an incubation period. I don't struggle with my mind under tension, but hope that a good creative idea will burst forth, and often it does. It sometimes is useful to write the problem and ideas in words. It is clear that much of the creative process is done subconsciously. But, if a good idea suddenly bursts forth, one is then involved in intense mental concentration, which may be quite prolonged.

It is important to distinguish sharply between two fundamental mental attributes, intelligence and imagination. We are all familiar with our judgement of intelligence in others using such attributes as quickness of grasp, depth of understanding, clarity of expression, range of intellectual interests, and especially insight. It can be measured by tests and, on the doubtful grounds that it can be assessed as a one-dimensional function, one is given an IQ value. Imagination is a much more subtle mental phenomenon, and I know of no test to evaluate it. Yet creative imagination is a property of the brain and mind that is of paramount importance. Imagination cannot be learnt. It is an endowment that we can be overwhelmingly grateful for. That is my personal attitude to my own endowment which has enabled me to do whatever I have done. And it does not seem to wither with age!

II. PROSPECTS

The prospects in neuroscience are illimitable. I have selected a few of the many attractive prospects in a somewhat arbitrary way. In part, my selection is related to principles that are apt to be overlooked or that are relatively new. I am aware that many important fields of investigation have been overlooked in this necessarily brief survey. For example, the immense fields of the sensory and motor systems would each require several lectures. Brooks (1986) has just published a valuable survey of the motor system.

1. The Building of the Brain

Primates are exceptional mammals in neurogenesis. Studies with tritiated thymidine show that there is no significant neurogenesis after birth. In fact, Rakic (1985) reports that the neurones constituting the neocortex of the rhesus monkey are all generated and have completed their migration during the first half of pregnancy; however, the synaptic connectivity has yet to be established (Rakic, 1981).

Rakic (1972) recognized the property of glia as forming the structural framework which guides the growth of the neuronal sprouts. This guidance has now to be considered in relation to the recent discoveries of cell adhesive molecules (CAM) (Edelman, 1984). A special class of these sialoglycoproteins encrust about 1% of the surface membranes of embryonic nerve cells. These embryonic CAM's have molecular weights of about 200,000 daltons, with a high concentration of sialic acid (about 30%). Furthermore, the adhesion to glia for guidance is attributable to a special type of N-CAM called Ng-CAM of 135,000 daltons. All CAM's project from the neuronal surface in searching for guidance both from glia and other nerve sprouts. These discoveries are of great importance in that they will probably provide explanations for the selectivities of neuronal growth which are of fundamental importance in establishing the precise neuronal connectivities of the brain.

The other important factor in determining the neuronal structure of the brain is selective neuronal death, which has been proposed by Cowan *et al.* (1984) as providing modelling and refinement of neuronal connectivity, which is rather analogous to sculpturing. Since in the primate no neuronal generation occurs after mid-pregnancy, a good biological strategy would be to generate an excess of neurones and then fashion the neuronal structure by death of redundant neurones. Cowan *et al.* (1984) suggest that deficiencies of trophic substances may be the factor, as will be considered in the next section.

2. *Trophic Factors in the Central Nervous System*

The dependence of neurones on trophic factors is recognized by the reactions of neurones to lesions that interrupt connections either *to* target organs (retrograde transport from other neurones or muscle) or *from* other neurones (anterograde deprivation). Study of this trophic dependence has revealed that in key cases it is dependent on special polypeptides that have the general name, neuronotrophic factors (NTF), of which nerve growth factor (NGF) is a well-known example of sympathetic ganglion cells. Deprivation of retrograde transport of NTF results in the long known chromatolytic reaction of neurones (Eccles,

1986b) that leads on to death if the deprivation is severe enough and permanent. The other side of the coin is that neurones are generators, in some way not understood, of the NTF's that they donate to the synaptic boutons on their surface, thereby ensuring the maintenance of these boutons and the neurones to which they belong. Glia may also be a source of NTF's.

Well-designed experiments indicate that there is a considerable regenerative capacity in several regions of the mammalian brain, even in adults. Optimistically, one can predict that we are only at the beginning of an enterprise in which various surgical procedures, plus rehabilitation therapy and local administration of neuronotrophic factors, will be able to reduce some of the disabilities suffered by patients with lesions of the brain and spinal cord. Regeneration has already been demonstrated in many regions of the mammalian brain: the septal nuclei, the red nucleus, the hippocampus, the lateral geniculate nucleus, the superior colliculus, the cerebellothalamic tract, and the cerebral cortex.

The most complete study has been made by Tsukahara (1981) and his associates at Osaka in their studies on the synaptic replacement and development of red nucleus neurones. The work was remarkable in that histological evidence was correlated to precise intracellular studies of the excitatory postsynaptic potentials (EPSPs) produced by the regenerating fibre systems. For the most part there was replacement of the degenerating synapses, as had been so carefully studied by Raisman and his colleagues in the septal nucleus. In addition, Tsukahara and associates also discovered that there was synaptic development of red nucleus neurones during recovery from a cross-union of the nerves to extensors and flexors of the kitten forelimb, and even during a conditional learning experiment.

All of these synaptic regenerations could be very localized, the nerve sprouts growing across a space of no more than 50 μm. In striking contrast is the regeneration discovered by Kawaguchi and associates in Kyoto (1986). After complete section of a major tract, the brachium conjunctivum (the cerebellothalamic tract) at its decussation, there was regeneration by outgrowths from the severed fibres for several millimeters so that effective synaptic connections were re-established in the thalamus. As shown by HRP injections into the cerebellar

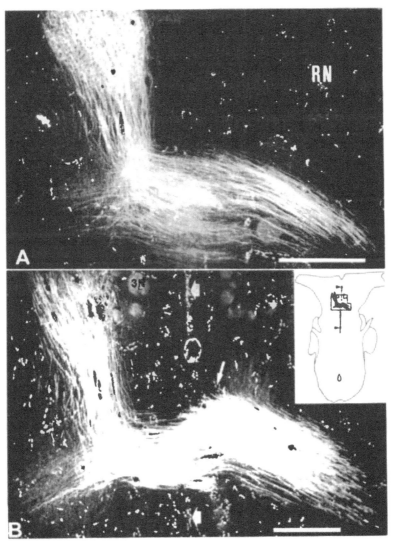

Fig. 1. A: dark-field photomicrographs of the decussation of the superior cerebellar peduncle in a horizontal section of a 6-day-old kitten labeled with HRP injected into the cerebellar nuclei on the right side. RN, red nucleus. B: as in A, but showing regenerated fibres after complete transection of the decussation of the brachium conjunctivum in a kitten 6 days old and prepared 19 days later. Glial scars in the lesion rostral and caudal to the area of fibre crossing are shown by the thick arrows. The thin arrow indicates some fibres deviating to take an ipsilateral course. Inset: the two arrows show the two holes left by ventral extraction of the vertical arms of the cutting device before the histological preparation. A-B scale bar: 500 μm. (Kawaguchi *et al.*, 1986)

nuclei (interpositus and lateral), there was a remarkable regeneration in favourable cases (Fig. 1). Unfortunately this regeneration occurred in only about 10% (8 out of 82 kittens and 3 out of 25 young adults), apparently being blocked by gliosis. The growth cones of the regenerating fibres appear to follow the degenerating fibres just as occurs after peripheral nerve section. The low success rate must be regarded as a challenge to investigate further the factors such as NTF's concerned in the regeneration and also to discover the cause of the low success rate. The important discovery is that under favourable conditions this amazing regeneration does occur. Thus there begins a new era of investigation of regeneration in the mammalian central nervous system. There has long been a belief dating from Ramon y Cajal that the embryonic know-how in development is lost. This inherited dogma has to be relinquished and we can have new hope for regenerative recovery from many lesions of the human central nervous system. That is a grand prospect.

3. Transplantation (Björklund and Stenevi, 1984)

It is extraordinary that the brain is "immunologically privileged" in that injections of neural or other tissue into the brain are usually not rejected, as they are when injected into other organs. One explanation of this privilege is that the brain lacks lymphatics and lymph nodes that are sources of many cells of the immune system, and also that the brain is protected by the blood-brain barrier. The great advantage is that there is opened up a wide range of opportunities for supplementing defective or inadequate neuronal tissue, which is the cause of such diseases as Parkinsonism and Alzheimer's disease. Transplanted nerve fragments from the embryo or young animals survive for long periods and establish appropriate neural connections as in the hippocampus. Embryonic neurones can survive when injected as a dissociated cell suspension. A general finding is that a transplant does better in a denervated region and that it is aided by the local neuronal damage caused by making a cavity for its reception, particularly if it is done some time before the transplantation. Presumably these conditions cause the production of the essential NTF's.

Very special interest attaches to the transplantation of a sympathetic ganglion into the hippocampus (Björklund and Stenevi, 1981) because for the first time there is a clear demonstration that a growth factor is released in the brain in quantities to cause very strong growth of a transplant. An essential prerequisite is that the septohippocampal (cholinergic) pathway be severed. A sympathetic ganglion transplanted into the hippocampus of an adult rat then hypertrophies, both in cell size and in fibre output, much as would occur peripherally under the influence of NGF. Furthermore, if there is axotomy of the locus coeruleus neurons innervating the hippocampus by intraventricular injection of a specific toxic agent, the regeneration of these adrenergic axons is greatly increased. It is postulated that the septal pathway normally suppresses the secretion of the neuronotrophic factors by the astroglia. There is evidence that astroglia in the hippocampus secrete neuronotrophic factors for both adrenergic and cholinergic nerve fibres. The special feature is that this secretion is assumed to be under inhibitory control by the septal input to the hippocampus. This leads to the hope that the poor regenerating ability of the mammalian brain may be attributable in part to as yet unknown controls of the secretion of neuronotrophic factors that facilitate regeneration.

Björklund and Stenevi (1979) have concentrated on monoaminoergic and cholinergic transplants into the hippocampus, because they can be easily monitored biochemically. Transplants into the hippocampus were from 16 to 17-day-old rat fetuses and were from the septal diagonal band (cholinergic), the locus coeruleus (noradrenergic), the raphe region (serotonergic), and the ventral mesencephalon (dopaminergic). It was found that fibres from the transplants grew well and closely mimicked the normal distribution, which is well known in the hippocampus. There is preferential growth into denervated hippocampal regions. It would seem that the new fibre growths are attracted to the filling of vacant fibre pathways and endings. It may be assumed that, since the ingrowing nerve terminals receive from their neurones the normal enzyme systems for transmitter manufacture, they are functionally effective. The transplants survive for over 1 year and provide a model for successful transplantation.

Implanted dopamine neurones grow preferentially into the corpus striatum, which is the normal target for the dopaminergic fibres from the substantia nigra (Dunnett et al., 1983). In an experimental model of Parkinsonism the movement disorders were reduced after the grafts.

These findings are encouraging, but much has to be accomplished before there is an effective treatment of Parkinsonism. Similarly, Alzheimer's disease may be treatable by transplants. Attempts have been made with some success to inject into the hippocampus embryonic precursor cells of the cholinergic medial septal area in order to treat rats suffering from the memory defects arising from a lesion of the septal cholinergic pathway to the hippocampus.

4. The Neurone

The nucleus plays a key role in the building of the structure of the neurone with its dendrites and axon (Fig. 2). But the manner in which it accomplishes this immense and intricate task is almost unknown. The DNA metabolism is unique because except for special limited sites, neurones have lost the ability to subdivide.

A special feature of the neurone is the axonal and dendritic transport along microtubules, which is shown in longitudinal and transverse section in Fig. 2. Ochs (1983) has presented an attractive model of the mechanism of the transport in which the particles to be transported are attached to "rafts" which are propelled by a lashing movement of cilia-like projections from microtubules. This model needs rigorous testing to see how far it can account for both retrograde and anterograde transport along the same microtubule and at a wide range of velocities from the maximum of 400 mm/day. Large particles such as mitochondria travel slowly since they seem to be repeatedly offloaded. As we have seen, axon transport in both retrograde and anterograde directions is vitally concerned in the trophic mechanisms.

The surface membrane is of particular importance for the neurone. Figure 3A shows the membrane as a typical bimolecular leaflet with various polypeptide molecules located transversely on it. Figure 3B illustrates three important functions of these "organelles" of transverse polypeptides. B1 and B2 are particularly related to synapses.

B1 depicts an ionic channel across the membrane. Under the influence of the synaptic transmitter substance this *ionotropic channel* opens momentarily for cations (sodium-potassium) or anions (chloride) and the ions run down their electrochemical gradients, so causing the excitatory or inhibitory postsynaptic potentials (EPSPs or IPSPs). In recent years the ionotropic channels have been studied by the patch-

These findings are encouraging, but much has to be accomplished before there is an effective treatment of Parkinsonism. Similarly, Alzheimer's disease may be treatable by transplants. Attempts have been made with some success to inject into the hippocampus embryonic precursor cells of the cholinergic medial septal area in order to treat rats suffering from the memory defects arising from a lesion of the septal cholinergic pathway to the hippocampus.

4. The Neurone

The nucleus plays a key role in the building of the structure of the neurone with its dendrites and axon (Fig. 2). But the manner in which it accomplishes this immense and intricate task is almost unknown. The DNA metabolism is unique because except for special limited sites, neurones have lost the ability to subdivide.

A special feature of the neurone is the axonal and dendritic transport along microtubules, which is shown in longitudinal and transverse section in Fig. 2. Ochs (1983) has presented an attractive model of the mechanism of the transport in which the particles to be transported are attached to "rafts" which are propelled by a lashing movement of cilia-like projections from microtubules. This model needs rigorous testing to see how far it can account for both retrograde and anterograde transport along the same microtubule and at a wide range of velocities from the maximum of 400 mm/day. Large particles such as mitochondria travel slowly since they seem to be repeatedly offloaded. As we have seen, axon transport in both retrograde and anterograde directions is vitally concerned in the trophic mechanisms.

The surface membrane is of particular importance for the neurone. Figure 3A shows the membrane as a typical bimolecular leaflet with various polypeptide molecules located transversely on it. Figure 3B illustrates three important functions of these "organelles" of transverse polypeptides. B1 and B2 are particularly related to synapses.

B1 depicts an ionic channel across the membrane. Under the influence of the synaptic transmitter substance this *ionotropic channel* opens momentarily for cations (sodium-potassium) or anions (chloride) and the ions run down their electrochemical gradients, so causing the excitatory or inhibitory postsynaptic potentials (EPSPs or IPSPs). In recent years the ionotropic channels have been studied by the patch-

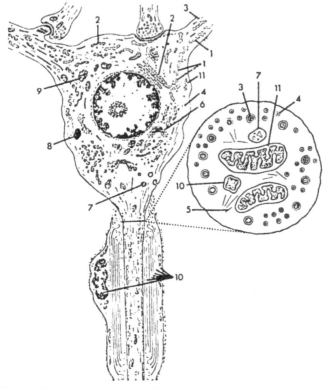

Fig. 2. Schematic diagram showing many features of a neuron as seen in electron micrographs. Shown are two dendritic processes and a myelinated axon (also in cross section) wrapped by processes from an oligodendroglial cell. An axodendritic and an axosomatic synapse are highlighted. Within the neuron are shown the subcellular organelles. These include the rough (2) and smooth (1) endoplasmic reticulum, the Golgi apparatus with secretory vesicles (6), lysosomes (8), lipofuscin granules (9), multivesicular bodies (7), microtubules (3), neurofilaments (4), microfilaments (5), and ribosomes. The nucleolus is shown within the nucleus. (10) oligodendroglial cell; (11) mitochondrion. (McGeer *et al.*, 1987)

clamp (Corey, 1983) and this exquisite technique offers most attractive prospects for future research. Furthermore, their molecular configuration is now being disclosed as an amazingly organized structure of polypeptides. It is a highly specialized field of organic chemistry.

B2 depicts a metabotropic structure in which the transmitter activates a protein kinase that in turn activates a second messenger system that works specifically on metabolic processes on the inner side

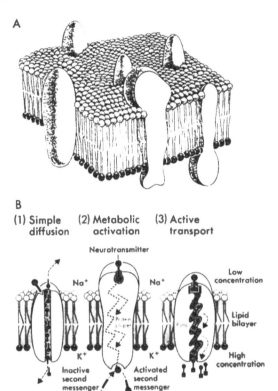

Fig. 3. A: organization of the surface membrane. The basic structure is a bimolecular leaflet of phospholipid molecules with hydrophilic structures on the external surface. One set is oriented to the extracellular fluid (stippled spheres) while the other is directed toward the intracellular fluid (black spheres). Structural proteins (not shown) are applied to both surfaces to provide stability. Six specific glycoprotein complexes are shown "dissolved" in the membrane. B: diagrammatic depictions of three types of receptors: (1) an ionotropic receptor that opens ion channels on activation by a neurotransmitter; (2) a metabotropic receptor that triggers a transmembrane reaction to a second messenger on activation by a neurotransmitter; and (3) a high-affinity pumping system that may function in neurotransmitter uptake. In the text reference is made to similarly oriented pumping systems for ions. (McGeer *et al.*, 1987)

of the membrane, as is diagrammatically shown in Fig. 4. There is a wide range of neurotransmitter actions by metabotropic structures and a great field of neurochemical investigation is open.

B3 depicts yet another structure across the membrane where metabolic energy is used to transport molecules or ions across the membrane. The best known example is the linked sodium-potassium

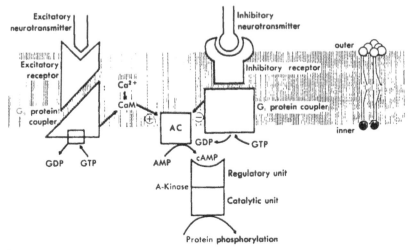

Fig. 4. Proposed mechanism of action of the cAMP second-messenger system. An excitatory or an inhibitory neurotransmitter interacts with its own receptor, which then couples to its own stimulatory (G_s) or inhibitory (G_i) protein. The consequence is to displace an inactive GDP molecule with an active GTP molecule. In the case of the G_s protein, there is activation of a Ca^{2+}-calmodulin complex (CaM), which in turn stimulates adenylate cyclase (AC) to produce cAMP. Cyclic AMP binds to the regulatory unit of A-kinase, releasing the catalytic unit to phosphorylate proteins. On the other hand, when the G_i protein is activated by GTP, it inhibits AC. (McGeer *et al.*, 1987)

pump. But there are also chloride and calcium pumps. As yet these ionic pumps have been fully investigated in only a few very specialized membranes as, for example, the squid giant axon.

5. The Organization of the Brain

There is no doubt that the field of neuronal connectivities presents an enormous prospect. Tracer techniques have been developed utilizing radio-tracers, immunohistochemistry, dyes as tracers, and macromolecules such as horseradish peroxidase that are transported in both directions along nerve fibres. And, of course, there are the techniques involving nerve degenerations. Somogyi *et al.* (1979) developed a combination of technical procedures: Golgi staining, retrograde transport of horseradish peroxidase and anterograde degeneration of synaptic boutons—in order to disclose the origin of input lines to a particular neurone and also the distribution of its axonal branches to other neurones.

The field of research into neuronal connectivity is immense. It is important to adopt rather rigorous criteria in controlling and evaluating the problems to be investigated. As mentioned under Strategy the ideal method is to develop specific hypotheses that can be experimentally tested. Critical evaluation must be applied to all connectivities disclosed by the extremely sensitive tracer techniques, because even single fibre connections are revealed. Already the described connectivities cannot in many cases be assimilated into a functional system. I think, for example, of the supplementary motor area. It has to be recognized that many observed connectivities may merely be aberrancies that were not rejected in the initial building of the brain, and which do not appreciably disturb the integrative performance of the neural systems on the brain.

6. Synapses

Synapses may be defined as structures subserving transmission between one neurone and another. In the brains of higher animals this transmission is chemical by specialized substances. Electric synaptic transmission is negligible. With electron microscopy the characteristic feature of a synapse is a swelling of the axon, called a bouton, filled with synaptic vesicles containing the synaptic transmitter substance and in close contact, across the synaptic cleft, with the postsynaptic membrane that has specialized receptors for the transmitter substance. Two boutons can be seen on the top of the neurone in Fig. 2. By special histological procedures it is possible to distinguish between the excitatory and inhibitory synapses. Besides these organized structures the term synapse should be extended to the rather ill-defined relationship between the varicosities of non-medullated fibres and their target organs on nerve cells or smooth muscle fibres (Fig. 5D).

Synapses are also distinguished by their synaptic transmitters that act in an ionotropic manner, glutamate and aspartate open Na^+-K^+ channels and are excitatory, while γ-amino butyric acid (GABA) and glycine open Cl^- channels and are inhibitory. Acetylcholine (ACh) rarely acts as an ionotropic excitant, as on Renshaw cells. Otherwise, ACh acts as a metabotropic transmitter in either an excitatory or inhibitory capacity (cf. Fig. 4). The amines, noradrenaline, serotonin,

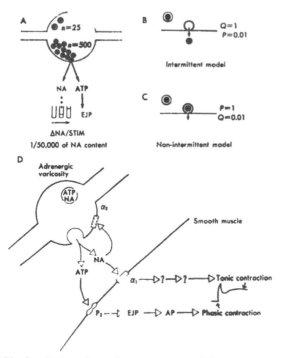

Fig. 5. A: in guinea pig or mouse vas deferens average varicosity contains some 500 small and 25 large densely—cored vesicles. Low-frequency stimulation preferentially induces secretion of transmitter (noradrenaline (NA) and ATP) from small vesicles. If every varicosity secreted the NA contained in 1 vesicle, output per stimulus (NA/stim) from tissue would be 1/500 of its NA content, but it is, in fact, 2 orders of magnitude less. Secretion of ATP can be monitored on an impulse-by-impulse basis by recording fast EJPs. B and C: intermittent and nonintermittent models, respectively. (Stjärne, 1986) D: schematic representation of the co-transmitter hypothesis proposed for guinea pig vas deferens. When the nerve varicosity is depolarized, it releases ATP and NA which act on P_1 and α_1-receptors, respectively, of the smooth muscle cell. The first phase of the contractile response results from the action of ATP depolarizing the cell and depends on the summation of EJPs' to fire action potentials (AP). The second phase is mediated by α_1-receptors by a mechanism which is independent of action potentials. (Sneddon and Westfall, 1984)

and dopamine are mostly acting out of varicosities and probably can be regarded as modulatory on the neurones.

Immunohistochemistry has been very effectively employed in identifying synaptic peptides. More than 20 are now known, but they are never alone. Concomitance is the rule. Often this is indicated histologically where the large dense core vesicles characteristic of the peptides occur with vesicles as in the varicosity of Fig. 5A. A useful classification of synaptic transmitters is as follows. Type 1 is the amino acids (glutamate, aspartate, GABA, and glycine); type 2 is the amines (ACh, dopamine, noradrenaline, serotonin, and histamine); type 3 is the peptides. Concomitance in synapses has been reported for all possible combinations of types. Most numerous are type 2 with type 3 and type 3 with type 3. It must be recognized that this concomitance is for the identified transmitters, and there are few examples as yet of concomitance in synaptic actions. This is most difficult to establish since most of the transmitters of types 2 and 3 probably function in an ill-defined modulatory manner. There are great prospects in all of this synaptic transmitter field.

The clearest example of concomitance of synaptic action has been observed by Stjärne (1986) and associates in their study of the secretion of transmitters from sympathetic nerve varicosities in the vas deferens. As illustrated in Fig. 5A there are in a varicosity about 500 small vesicles containing noradrenaline (NA) and ATP. With nerve stimulation and intracellular recording from a smooth muscle fibre, there is an excitatory junction potential (EJP) that is due to ATP acting in a fast ionotropic manner. NA acts metabotropically to give a much smaller and slower muscle contraction as shown in Fig. 5D. The amount of NA liberated by a nerve stimulation is about 1% of that expected if there was the liberation of one vesicle per impulse from each varicosity. Correspondingly, the EJP's are only rarely produced by emission of ATP from a varicosity. By analysis, Stjärne and associates have shown (Fig. 5B) that low probability of vesicle emission occurs for the EJP's produced by ATP (1%) as also for the NA emission. Hence the diagram of intermittency is drawn as in Fig. 5B, and also in Fig. 5D. This low probability of emission (1%) is a necessary conservation, otherwise varicosities would quickly be depleted. The large dense core vesicles of Fig. 5A are difficult to study, but they are not linked in emission with the small vesicles. The NA and ATP actions in Fig. 5D have been identified by selective blocking agents (Sneddon and Westfall, 1984).

As Stjärne (1986) points out, this concept of a very low probability of vesicular emission from varicosities is a complete transformation

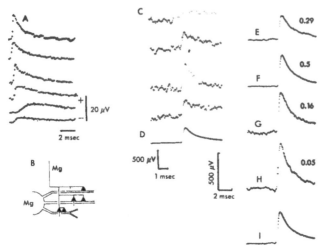

Fig. 6. A: averaged recordings of EPSPs produced by impulses in the same Ia fibre terminating on six different motoneurons. (Mendell and Henneman, 1971) B: summary diagram of the location of Ia synapses from a single medial gastrocnemius Ia fibre on a medial gastrocnemius motoneuron at five sites on three different dendrites as indicated. (Brown, 1981) C: four individual EPSPs selected from a population of 800 responses. D: the average of all the 800 responses. E: component 1 of the EPSP derived from fluctuation analysis. F–H: components 2, 3, and 4 of this same fluctuation analysis. The probabilities of the occurrence of these components are indicated to the right of each. I: the reconstructed EPSP obtained by adding the weighted sum of E, F, G, H; $0.29E+0.5F+0.16G+0.05H$. (Jack et al., 1981)

of the classical story. There is thus opened up an exciting new prospect in research. Furthermore, it relates to recent concepts on the probability of vesicle emission from boutons of the somatic system that will now be described.

The ultimate refinement of synaptic investigation derives from the studies of the manner in which a presynaptic impulse causes exocytosis of synaptic vesicles with a complete emptying of their contained transmitter molecules. When the synaptic input to a motoneurone is cut down to a single group Ia fibre with its several boutons (Fig. 6B), there is a wide range in the successive unitary EPSPs (Fig. 6C). This is attributable to variations in the bouton contributions, each bouton giving a unitary vesicular output every now and then. By a very sophisticated technique of fluctuation analysis Redman and associates Jack et al., 1981) have been able to determine the contribution of individual boutons, there being in the illustrated case 4 such quantal

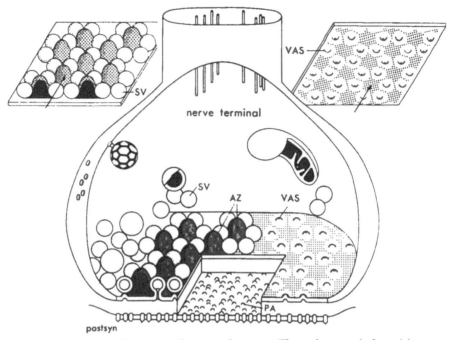

nerve terminal

Fig. 7. Schema of the mammalian central synapse. The active zone is formed by pre-synaptic dense projections (AZ). The postsynaptic aggregation of intramembranous parti-cles is restricted to the area facing the active zone, SV, synaptic vesicles; PA, particle aggregations on postsynaptic membrane (postsyn.). Note synaptic vesicles (SV) in hexago-nal array, as is well seen in the upper left inset, and the vesicle attachment sites (VAS) in the right inset. (Akert *et al.*, 1975)

emitters, with probabilities shown in Figs. 6E to H. It turns out that the probability of vesicular exocytosis is less than one for a bouton, with usual values of about 0.3. A similar range of probabilities has been found in other species of synapses, notably by Korn and Faber (1987) with the inhibitory synapses on the Mauthner cell of the gold-fish. So, here we have an important new synaptic property that should be studied for other systems.

In attempting to understand the synaptic mechanism underlying this probability, we turn to the idealized bouton drawn by Akert *et al.* (1975) as a result of electron microscopic and freeze fracture tech-niques (Fig. 7). Each bouton confronts the synaptic cleft by a para-crystalline structure composed of presynaptic dense projections and synaptic vesicles in hexagonal array in close contact with the presynaptic membrane fronting the synaptic cleft. Evidently the presynaptic vesicular grid must work in a global manner conserving vesicles so that an impulse causes a single exocytosis with a low probability in accord with the fluctuation analysis of the unitary EPSPs in Fig. 6C.

The manner of operation of the presynaptic vesicular grid in controlling vesicular emission is completely unknown. Already factors are known that increase or decrease the probability. The probability of operation of boutons opens up entirely new conceptual fields in the cerebral cortex, such as consciousness (Eccles, 1986a) and memory (Eccles, 1987).

7. Learning and Memory

It can be argued that the most important function of the brain is memory. There is an enormous volume of investigations in animals and man on experimental testing of memory, for example, in the study of operant conditioning of animals and the effect of a wide variety of cerebral lesions on it. Unfortunately all of these investigations remain at the descriptive level unless they are related to the testing of some specific hypothesis. Attempts are now being made to locate the site of the engram, as Lashley called it. For example, Ito and associates (Ito, 1984) locate the learning processes of the vestibulo-ocular reflex in the cerebellar flocculus, the climbing fibre error detection there modifying the responses of the mossy fibres from the vestibular apparatus. In a notable study of motor learning in monkeys Sasaki and Gemba (1983) have shown the gradual development of enhanced responses of many cortical areas during the learning task, but the smooth skilled movement does not develop until the cerebellum participates in cerebro-cerebellar learnt responses. However, there is still no study of the synaptic mechanisms involved in the learning. Thompson (1987) would locate the engram for the conditioned eye blink reflex in the interpositus nucleus of the cerebellum, but I think this nucleus is functioning as a relay from an unrecognized learning area in the cerebellar cortex, for which there is now some evidence.

The outstanding problem with memory is to discover the synaptic basis for the long enduring changes involved in memory, that may persist for hours, days, weeks, or years. Ever since the hippocampal study by Bliss and Lømo (1973) the hippocampus has been studied

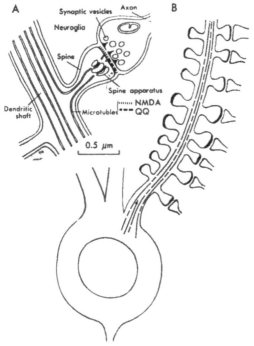

Fig. 8. A: drawing of spine synapse of a dendrite of a hippocampal pyramidal cell. From the presynaptic membrane there are dense projections up to and between the synaptic vesicles. The postsynaptic density is shown with two distinct receptor sites for NMDA and QQ as described in the text and as labelled below the diagram (modified from Gray, 1982). B: drawing of a dentate granule cell showing dendrites in outline with synaptic spines drawn on both sides of it. Microtubules are drawn in the right dendrite by three interrupted lines, and arrow shows lines of transport.

intensively as a memory model. Figure 8A shows diagrammatically the hypothesis of the synaptic mechanism of long term potentiation (LTP) on the hippocampus. The hypothesis is that: the postsynaptic receptor of the spine synapse is a double receptor structure. There is much recent evidence that a single synaptic bouton of a spine synapse has two distinct receptor sites for the transmitter, glutamate. One is the conventional quisqualate (QQ) receptor that, when activated, opens channels to Na^+ and K^+ ions, so producing the large depolarization of the EPSP. The other is specially excited by N-methyl-d-aspartate (NMDA) and blocked by 2 amino-5-phosphonovalerate (APV). Activation of this NMDA receptor by glutamate gives normally a negligible depolarization. However, when the extracellular Mg^{2+} concentration is greatly reduced or when the receptor is subjected to a steady large depolarization, glutamate activated NMDA receptors give a large slow depolarization of the spine lasting for about 60 msec with a large influx of Ca^{2+} ions through the opened Ca^{2+} channels (Gustafsson and Wigström, 1986). APV prevents this depolarization of NMDA receptors and the opening of the Ca^{2+} channels. As would be expected, LTP is also blocked by APV.

In the normal operation of a cortical neurone such as that partly shown in Fig. 8B, there would be activation of hundreds of synapses with the current passing through the QQ receptors to give a large EPSP. Furthermore, this cooperatively would produce sufficient depolarization for opening the Ca^{2+} channels of those NMDA receptors activated by glutamate at that time. This correlates with the finding that only those previously activated synapses exhibit LTP. These synapses with double receptor sites provide an effective model for memory and a great challenge for further study. Especially notable are the precise analytic studies by Gustafsson and Wigström (1986) on NMDA receptors for glutamate which may well be the key structures in the laying down of memories in the cerebral cortex.

8. Technical Procedures with Great Promise

Small fragments of the mammalian brain can survive for many hours or even days under stringent conditions that allow oxygenation and a supply of nutrients and other essential substances. The most developed technique is the tissue slice. Thin slices of less than 0.4 mm thickness are rapidly cut from the chosen region of the brain and transferred to the oxygenated medium in a few minutes. Necessarily there is great damage to many neurons, but it is hoped that in an hour or so these damaged regions have ceased to disturb the intact neurons of the slice which can then be studied under direct vision. By orienting the section it is possible to stimulate the input and observe the output, and, of course, intracellular recording is greatly aided in comparison with the *in vivo* preparation. The disadvantages are the unknown damage inflicted on apparently intact neurones by the section, and the short duration, a few hours, before irreversible deterioration sets in.

Cultures of fragments of embryonic brain have promise for the future. Neurones mature and establish functional synaptic connections. The cellular monolayer culture has the advantage over the slice in that there is no complication by tissue damage, and survival is for weeks (Gähwiler, 1984).

In quite a different logistic strategy there are the remarkable studies of the whole human brain by sophisticated techniques of PET scanning, nuclear magnetic resonance, or radio Xenon. With progressive improvement in the spatial and temporal grain of the procedures, most valuable data are obtained from human brains of conscious subjects doing prescribed tasks. We are in a new era of studying the human brain.

REFERENCES

Akert, K., Peper, K., and Sandri, C. (1975). Structural organization of motor end plate and central synapses. In *Cholinergic Mechanisms*, ed. Waser, P.G., pp. 43–57, New York: Raven Press.

Björklund, A. and Stenevi, U. (1979). Regeneration of monoaminergic and cholinergic neurons in the mammalian central nervous system. *Physiol. Rev.* **59**, 62–100.

Björklund, A. and Stenevi, U. (1981). *In vivo* evidence for a hippocampal adrenergic neuronotrophic factor specifically released on septal deafferentation. *Brain Res.* **229**, 403–428.

Björklund, A. and Stenevi, U. (1984). Intracellular neural implants: neuronal replacement and reconstructing of damaged circuitries. *Annu. Rev. Neurosci.* **7**, 279–308.

Bliss, T.V.R. and Lømo, T. (1973). Long-lasting potentiation of synaptic transmission in the dentate area of the anaesthetized rabbit following stimulation of the perforant path. *J. Physiol.* **232**, 331–356.

Brooks, V.B. (1986). *The Neural Basis of Motor Control*. New York, Oxford: Oxford Univ. Press.

Brown, A.G. (1981). *Organization in the Spinal Cord: The Anatomy and Physiology of Identified Neurones*. Berlin, Heidelberg, New York: Springer-Verlag.

Corey, D.P. (1983). Patch clamp: Current excitement in membrane physiology. *Neurosci. Commun.* **1**, 99–110.

Cowan, W.M., Fawcett, J.W., O'Leary, D.D.M., and Stanfield, B.B. (1984). Regressive effects in neurogenesis. *Science* **225**, 1258–1265.

Dunnett, S.B., Björklund, A., Schmidt, R.H., Stenevi, U., and Iversen, S.D. (1983). Intracerebral grafting of neuronal cell suspensions. V. Behavioural recovery in rats with bilateral 6-OHDA lesions following implantation of nigral cell suspensions. *Acta Physiol. Scand.* (Suppl.) **522**, 39–47.

Eccles, J.C. (1986a). Do mental events cause neural events analogously to the probability fields of quantum mechanics? *Proc. Roy. Soc. Lond. B* **227**, 411–428.

Eccles, J.C. (1986b). Chromatolysis of neurons after axon section. In *Recent Advances in Restorative Neurology*, vol. 2. Progressive neuromuscular diseases. eds. Dimitrijevic, M.R., Kakulas, B., and Vrbova, G., pp. 318–331. Basel: Karger.

Eccles, J.C. (1987). Mammalian systems for storing and retrieving information. In *Cellular*

Mechanisms of Conditioning and Behavioring Plasticity, ed. Woody, C.D. New York: Plenum Press (in press).

Edelman, G.M. (1984). Modulation of cell adhesion during induction, histogenesis and prenatal development of the nervous system. *Annu. Rev. Neurosci.* **7**, 339–377.

Gähwiler, B.H. (1984). Slice cultures of cerebellar, hippocampal and hypothalamic tissue. *Experientia* **40**, 235–243.

Gray, E.G. (1982). Rehabilitating the dendritic spine. *Trends Neurosci.* **5**, 5–6.

Gustafsson, B. and Wigström, H. (1986). Hippocampal long-lasting potentiation produced by pairing single volleys and brief conditioning tetani evoked in separate afferents. *J. Neurosci.* **6**, 1575–1582.

Ito, M. (1984). *The Cerebellum and Neural Control.* New York: Raven Press.

Jack, J.J.B., Redman, S.J., and Wong, K. (1981). The components of synaptic potentials evoked in cat spinal motoneurones by impulses in single group Ia afferents. *J. Physiol.* **321**, 65–96.

Kawaguchi, S., Miyata, H., and Kato, N. (1986). Regeneration in peduncle in kittens: Morphological and electrophysiological studies. *J. Comp. Neurol.* **215**, 258–273.

Korn, H. and Faber, D.S. (1987). Regulation and significance of probabilistic release mechanisms at central synapses. In *New Insights into Synaptic Function*, eds. Edelman, G.M., Gall, W.E., and Cowan, W.M., pp. 57–108, New York: Neurosciences Research Foundation, New York: John Wiley.

McGeer, P., Eccles, J.C., and McGeer, E. (1987). *Molecular Neurobiology of the Mammalian Brain*, 2nd ed. New York: Plenum Press.

Mendell, L.M. and Henneman, E. (1971). Terminals of single Ia fibers: location, density and distribution within a pool of 300 homonymous motoneurons. *J. Neurophysiol.* **34**, 171–187.

Ochs, S. (1983). Axoplasmic transport. In *Handbook of Neurochemistry*, vol. 5, ed. Lajtha, A., pp. 355–379. New York: Plenum Press.

Rakic, P. (1972). Mode of cell migration to the superficial layers of fetal monkey neocortex. *J. Comp. Neurol.* **145**, 61–84.

Rakic, P. (1981). Developmental events leading to laminar and areal organization of the neocortex. In *Organization of the Cerebral Cortex.* eds. Schmidt, F.O., Worden, F.G., Edelman, G., and Dennis, S.G., pp. 7–28. Cambridge, Mass.: MIT Press.

Rakic, P. (1985). DNA synthesis and cell division in the adult primate brain. *Ann. N.Y. Acad. Sci.* **457**, 193–211.

Sasaki, K. and Gemba, H. (1983). Learning of fast and stable hand movement and cerebro-cerebellar interactions in the monkey. *Brain Res.* **277**, 41–46.

Sneddon, P. and Westfall, D.P. (1984). Pharmacological evidence that adenosine triphosphate and noradrenaline are co-transmitters in the guinea-pig vas deferens. *J. Physiol.* **347**, 561–580.

Somogyi, P., Hodgson, A.J., and Smith, A.D. (1979). An approach to tracing neuron networks in the cerebral cortex and basal ganglia: Combination of Golgi-staining, retrograde transport of horseradish peroxidase and anterograde degeneration of synaptic boutons in the same material. *Neuroscience* **4**, 1804–1852.

Stjärne, L. (1986). New paradigm: Sympathetic neurotransmission by lateral interaction between secretory units? *News Physiol. Sci.* **1**, 103–106.

Thompson, R.F. (1987). Identification of the essential memory trace circuit for a basic form of associative learning. In *Cellular Mechanisms of Conditioning and Behavioral Plasticity.* ed. Woody, C.D. New York: Plenum Press (in press)

Tsukahara, N. (1981). Synaptic plasticity in the mammalian central nervous system. *Annu. Rev. Neurosci.* **4**, 351–379.

2

THE ACETYLCHOLINE RECEPTOR: FUNCTIONAL ORGANIZATION AND EVOLUTION DURING SYNAPSE FORMATION

JEAN-PIERRE CHANGEUX,*1 ANNE DEVILLERS-THIÉRY,*1 JÉRÔME GIRAUDAT,*1 MICHAEL DENNIS,*1 THIERRY HEIDMANN,*1 FRÉDÉRIC REVAH,*1 CHRISTOPHE MULLE,*1 ODILE HEIDMANN,*1 ANDRÉ KLARSFELD,*1 BERTRAND FONTAINE,*1 RALPH LAUFER,*1 HOÀNG-OANH NGHIÊM,*1 EKATERINI KORDELI,*2 AND JEAN CARTAUD*2

*Unité de Neurobiologie Moléculaire et Unité Associée au Centre National de la Recherche Scientifique UA 041149, Interactions Moléculaires et Cellulaires, Département des Biotechnologies, Institut Pasteur, 75724 Paris Cédex 15, France*1 and Microscopie Électronique, Institut Jacques Monod du Centre National de la Recherche Scientifique, Université Paris VII, 75005 Paris, France*2*

Biological regulation at the molecular level involves specialized proteins which act as "biological transducers," recognize physiological signals, possess a biological activity, and mediate the interactions between signal binding sites and biologically active sites. In these "allosteric" proteins, exemplified by hemoglobin, regulatory enzymes or gene repressors and activators, such interactions have been shown to take place between topographically distinct sites (Changeux, 1961; Monod et al., 1963, 1965) and to be mediated by conformational transitions of the protein molecule, usually between a small number of discrete and symmetrical states (Monod et al., 1965; Krause et al., 1985; Perutz et al., 1986; Fischer and Olsen, 1986).

These concepts have been extended to the proteins involved in intercellular communication typified, in the nervous system, by the receptors for neurotransmitters (Changeux, 1966; Karlin, 1969; Changeux et al., 1967, 1984; Changeux, 1981). The neurotransmitter serves as regulatory signal and interacts with the recognition site of the receptor molecule which faces the outside of the cell; the biologically active site can either be a transmembrane ion channel (review Changeux et al., 1984; Hucho, 1986; Fischer and Olsen, 1986) or a protein binding domain whose interaction with regulatory proteins controls the activity of an enzyme catalytic site oriented toward the inside of the cell (Schramm and Selinger, 1984). Such molecules play a critical role in the transfer of information between cells at the level of the chemical synapses and may be subject to both short-term changes of their properties and longer-term regulation of their number.

In this paper, the work recently done in our laboratory on the nicotinic acetylcholine receptor (AChR) from *Torpedo marmorata* electric organ and vertebrate neuromuscular junction is briefly summarized and its relevance to plausible models for short and long-term changes of synapse properties discussed.

I. THE ACETYLCHOLINE NICOTINIC RECEPTOR: A *QUASI-SYMMETRICAL* TRANSMEMBRANE PROTEIN

The AChR light form is a glycoprotein of approx. 300 KD, MW (268 KD of protein) made up of four different polypeptide chains of exact molecular weight (in *Torpedo californica*) α (50,116), β (53,681), γ (56,279), and δ (57,565) (Noda et al., 1982, 1983a,b) assembled in a rather uncommon $2\alpha.1\beta.1\gamma.1\delta$ stoichiometry (Reynolds and Karlin, 1978; Lindstrom et al., 1979; Raftery et al., 1980; Saitoh et al., 1980).

The purified and membrane-bound AChR appears by electron microscopy on "en face" views as rosettes, 80 Å in diameter, with a central depression about 25 Å in width (Cartaud et al., 1973) and five peaks of density, regularly disposed around a 5-fold axis of symmetry (Bon et al., 1984; Brisson and Unwin, 1985), and, on the side, as a transmembrane cylinder (110 Å long) which extends above the lipid bilayer by 55 Å into the synaptic cleft and by about 15 Å into the cytoplasm (reviews Changeux, 1981; Kistler and Stroud, 1981; Wise et al., 1981; McCarthy et al., 1986; Hucho, 1986).

The cDNAs coding for the four chains have been cloned and sequenced in *T. californica* (Ballivet et al., 1982; Noda et al., 1982, 1983 a,b; Claudio et al., 1983), in *T. marmorata* (for the α chain: Giraudat et al., 1982; Sumikawa et al., 1982; Devillers-Thiéry et al., 1983), and in several vertebrate species, including human (see Noda *et al.*, 1983c; Kubo *et al.*, 1985; Boulter *et al.*, 1986). As initially noticed from amino terminal sequence analyses (Raftery *et al.*, 1980), the inferred amino acid sequences show striking homologies between subunits (from 10 to 60% identity with an average of 40%) (Noda *et al.*, 1983b) which support a *quasi*-symmetrical organization of the receptor oligomer.

Reconstitution experiments with the purified protein incorporated into lipid vesicles and planar lipid bilayers show that the $\alpha_2\beta\gamma\delta$ oligomer contains the ion channel (selective for Na^+, K^+, Ca^{2+}) and *all* the structural elements required for its opening by nicotinic agonists and for its regulation by allosteric effectors (review in Popot *et al.*, 1981; Montal *et al.*, 1986). Supporting this conclusion, injection of the four mRNAs coding for the receptor subunits into *Xenopus* oocytes yields functional acetylcholine-gated ion channels and, in *Torpedo*, the presence of the four subunits is required for a fully functional receptor (Mishina *et al.*, 1984).

The AChR is thus a compact *quasi*-symmetrical transmembrane channel protein to which regulatory signals have access from both the outside and the inside of the cell.

II. MODELS OF TRANSMEMBRANE TOPOLOGY OF THE SUBUNITS

The aligned sequences of the four AChR subunits show a similar hydrophobicity profile which justifies a common subdivision of the homologous chains into: 1) a *large hydrophilic* amino-terminal domain of 210–224 amino acids; 2) a compact *hydrophobic* region of 68 residues subdivided into three segments of 19–27 uncharged amino acids (numbered I, II, and III); 3) *a small hydrophilic* domain of 109–146 amino acids; 4) a carboxy-terminal segment of 20 *hydrophobic* residues (numbered IV). On the basis of these primary sequence data, *models* of transmembrane organization common to the homologous subunits have been proposed (Claudio *et al.*, 1983; Devillers-Thiéry *et al.*, 1983; Noda *et al.*, 1983b; Guy, 1984; Finer-Moore and Stroud, 1984; McCarthy *et al.*, 1986; Criado *et al.*, 1985; Ratnam *et al.*, 1986). Their common features are: 1) the orientation of the large hydrophilic domain towards the synaptic cleft; 2) the orientation of the small hydrophilic domain towards the cytoplasm; 3) the assignment of the four hydrophobic segments I to IV (only I to III for Ratnam *et al.*, 1986) to transmembrane α-helices. All assume that the acetylcholine binding site is located in the large hydrophilic domain on the α subunit. Differences concern, in particular, the number, orientation and identity of the transmembrane segment(s) contributing to the walls of the ionic channel, in all instances assumed to lie in the axis of quasi-symmetry of the receptor molecule and to be delineated by homologous portions of each subunit. In one group of models, each chain traverses the lipid bilayer four times and the carboxy terminal end faces the synaptic cleft, helix I (Noda *et al.*, 1983b) or III (Devillers-Thiéry *et al.*, 1983), being the transmembrane components of an *uncharged channel*. Another model postulates an additional fifth "amphipathic" helix, called A, formed at the expense of the cytoplasmic domain thus reorienting helix IV to expose the carboxy terminal to the cytoplasmic face. Helix A would then delineate a *charged channel* (Guy, 1984; Finer-Moore and Stroud, 1984; Mishina *et al.*, 1985). A recent model proposes that helix IV is not transmembrane but cytoplasmic and that a transmembrane "hairpin" loop forms from the large hydrophilic domain (from amino acid 142 to 192 on the α subunit); the segment 159–192 would then make an amphipathic helix appropriate for lining a charged channel (Ratnam *et al.*, 1986).

Tests for these models have been developed in several laboratories using different experimental approaches which include: 1) site directed mutagenesis and expression of cloned cDNAs for the α subunit (Mishina *et al.*, 1985); 2) immunochemical studies using antibodies directed against peptides of defined sequences (review Lindstrom, 1986); 3) covalent labeling of specific sites of the AChR with radioactive affinity (or photoaffinity) reagents and identification of the labeled peptides. The results obtained in our laboratory by this last method are summarized in the next two paragraphs.

III. LOCALIZATION OF THE ACETYLCHOLINE BINDING SITE IN THE LARGE NH$_2$-TERMINAL HYDROPHILIC DOMAIN

The $\alpha_2\beta\gamma\delta$ oligomer carries two primary acetylcholine sites to which nicotinic agonists, antagonists and snake venom α toxins bind in a reversible and mutually exclusive manner but which possess quite different binding properties (review Karlin, 1980; 1983; Changeux et $al.$, 1984). The two α subunits of the oligomer contribute to these sites, yet Southern blot hybridization of genomic DNA with specific cDNA probes supports the existence of a single gene coding for the α subunits in $Torpedo$ (Klarsfeld et $al.$, 1984) and in other species (Merlie, 1984; Klarsfeld and Changeux, 1985; Heidmann, O. et $al.$, 1986; references in McCarthy et $al.$, 1986). The different properties of the two acetylcholine sites may then result from differences in subunit environment within the oligomer and/or in posttranslational modifications.

Attempts to identify the amino acids which contribute to the acetylcholine binding site have been made on the basis of the observation that disulfide bond reduction perturbs the pharmacological response to agonists and permits the labeling of the α subunit by cholinergic affinity ligands specific for sulfhydryl groups (such as the maleimide reagents MBTA or MPTA) (review Karlin, 1969, 1983). Kao et $al.$ (1984) have demonstrated by peptide mapping and sequencing that the residues cystein 192 and possibly 193, a tandem unique to the α subunit, represent the sites of incorporation of MBTA, which, however, occur exclusively on the $reduced$ receptor.

In order to probe in greater detail the structure of the acetylcholine binding sites on the $native$ AChR, we used p-N,N-dimethylamino-, benzenediazonium fluoroborate (DDF) (Goeldner and Hirth, 1980) a derivative of TDF, a compound described as the first affinity labeling reagent of electric organ AChR (Changeux et $al.$, 1967a; Weiland et $al.$, 1979). In the dark DDF behaves as a competitive antagonist but it irreversibly blocks the binding of cholinergic ligands and snake venom α toxins following irradiation at 295 nm. Under appropriate conditions [^{3}H]DDF predominantly labels the α subunit of the native AChR and this labeling is prevented by cholinergic agonists and antagonists (Langenbuch-Cachat et $al.$, 1987). Comparative peptide mapping of the α subunit labeled by [^{3}H]MPTA and [^{3}H]DDF revealed differences in the distribution of the two markers. [^{3}H]DDF was incorporated into three distinct CNBr fragments in a carbamylcholine-sensitive manner. The predominant one corresponded to segment α 179–207, and was the only one labeled by [^{3}H]MPTA at the level of cysteins 192 and 193. In this fragment, [^{3}H]DDF labeled $tyrosine$ 190, $cysteins$ 192 and 193 and possibly subsequent amino acids (Dennis et $al.$, 1986 and unpublished data). This region is also involved in α-bungarotoxin binding (Wilson et al., 1985; Neumann et al., 1986). Two other fragments specifically labeled by [^{3}H]DDF were derived from a more amino terminal region of the large hydrophilic domain of the α subunit (unpublished data). Several segments of the large hydrophilic domain may thus participate in the binding of cholinergic ligands (see also Mulac-Jericevic and Atassi, 1986; McCormick and Atassi, 1984; but Criado et al., 1986). These results obtained with the native AChR are, thus, consistent with the above mentioned models which postulate that the large hydrophilic domain of the α subunit is at least in part exposed to the synaptic cleft and carries the acetylcholine binding site.

IV. THE HIGH AFFINITY SITE FOR "CHANNEL BLOCKERS"

In both the native postsynaptic membranes and reconstituted purified receptor (Popot et al., 1981), the ionic response to acetylcholine is inhibited by a class of particularly potent pharmacological agents, the noncompetitive blockers, which are postulated to interfere directly or indirectly (or both) with the ion channel (references in Neher and Steinbach, 1978; Adams, 1981). The binding site(s) at which these noncompetitive blockers exert their effect is different from the acetylcholine binding sites (Weber and Changeux, 1974) but strongly interacts with them under equilibrium conditions (Cohen et al., 1974; Grünhagen and Changeux, 1976; Eldefrawi et al., 1977; Krodel et al., 1979; review Heidmann et al., 1983). These sites have been subdivided into two main categories (Heidmann et al., 1983): 1) high affinity sites, sensitive to histrionicotoxin (Eldefrawi et al., 1977) and present as a unique copy per receptor pentamer; and 2) low affinity ones, insensitive to histrionicotoxin, much more numerous (10 to 20 times the number of α-bungarotoxin sites) and lipid-dependent, possibly located at the interface of the receptor with membrane lipids.

Early attempts to identify the chain (or chains) of the AChR involved in the high affinity site were done with T. marmorata receptor and an azido derivative of the potent local anesthetic trimethisoquin. Only the δ subunit incorporated the label in an agonist-dependent manner (Oswald et al., 1980; Oswald and Changeux, 1981a). Both [^{3}H]-labeled perhydrohistrionicotoxin and [^{3}H] phencyclidine under

UV irradiation (Oswald and Changeux, 1981b) gave similar results. However, with $T.$ *californica* and quinacrine mustard (Kaldany and Karlin, 1983), or triphenylmethylphosphonium (Muhn and Hucho, 1983) or with $T.$ *ocellata* and azidophencyclidine (Haring *et al.*, 1984), labeling occurred primarily at the level of the α or the β chains, respectively. With $T.$ *marmorata*, [^{3}H]-chlorpromazine labeled all four subunits after UV irradiation (Oswald and Changeux, 1981b).

Careful measurements confirmed the stoichiometry of one chlorpromazine high-affinity binding site per two α-toxin sites. Since, when bound to this site, chlorpromazine attaches covalently to any of the four chains of the AChR, it was proposed (Heidmann *et al.*, 1983) that this site lies in the *axis of quasi-symmetry* of the $\alpha_2\beta\gamma\delta$ oligomer where the distances to all five chains are minimum. Minor sequence differences between species of *Torpedo* and differences in ligand structure would then account for the various labeling patterns mentioned above.

Amino acids labeled by [^{3}H]-chlorpromazine were identified in this laboratory (Giraudat *et al.*, 1986, 1987). Photolabeling of the membrane bound receptor with [^{3}H]-chlorpromazine was carried out under equilibrium in the presence of carbamylcholine and in the absence, or presence, of phencyclidine, a selective ligand of the high-affinity site for noncompetitive blockers. The δ subunit was purified, digested with trypsin and the resulting fragments fractionated by reversed-phase high performance liquid chromatography (HPLC). The labeled peptide(s) could not be purified to homogeneity because of its (their) marked hydrophobic character. Yet, compared sequence analysis of the fraction of tryptic peptides containing the specifically labeled material before and after CNBr subcleavage and repurification lead to the unambiguous demonstration that Ser 262 is labeled by [^{3}H]-chlorpromazine. Sequence analysis of analogous material derived from control batches further showed that labeling of residue Ser 262 was prevented by phencyclidine. Residue Ser 262 which lies within the hydrophobic segment II of the δ subunit is thus part of (or in close vicinity to) the high-affinity site for noncompetitive blockers (Giraudat *et al.*, 1986). Similar experiments were repeated with the β chain and sequence analysis resulted in the identification of Ser 254 and Leu 257 as residues labeled by [^{3}H]-chlorpromazine in a phencyclidine-protectable manner (Giraudat *et al.*, 1987). These residues are also located

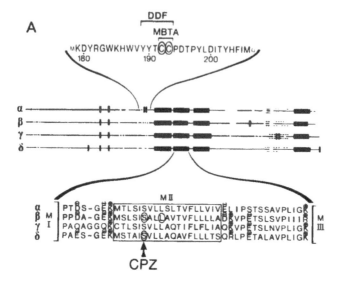

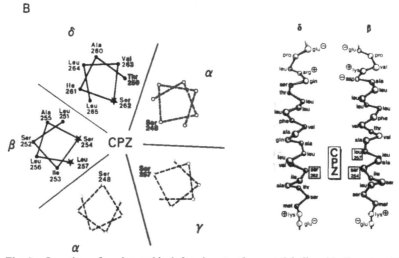

Fig. 1. Location of amino acids belonging to the acetylcholine binding site (DDF, MBTA) and to the high affinity site for the noncompetitive "channel" blocker chlorpromazine (CPZ) on the amino acid sequence of the four subunits of the AChR from *T. marmorata*. A: aligned sequences of the four subunits (from Changeux *et al.*, 1987). B: hypothetical model of the front and side views of the high affinity site for CPZ. From Giraudat *et al.* (1987).

in the hydrophobic and potentially transmembrane segment M II of the β-subunit, a region homologous to that containing chlorpromazine-labeled Ser 262 in the δ subunit.

Subsequently, using a different noncompetitive blocker [³H]-triphenylmethylphosphonium, Oberthür et al. (1986) have identified Ser 262 as a labeled amino acid on the δ subunit, a finding consistent with our original observation with [³H]-chlorpromazine. A preliminary report from the same group (Hucho et al., 1986) also provides suggestive evidence for a labeling on the α and β subunits at homologous positions.

Homologous regions of different receptor subunits thus contribute to the unique high affinity site for noncompetitive blockers, a finding consistent with the proposal (Heidmann et al., 1983) that this site is located in the axis of quasi-symmetry of the AChR molecule. Moreover, the segments M II are non homologous to the regions labeled by affinity reagents of the acetylcholine binding site on the α subunits. These structural data are thus consonant with the concept that the interactions between the acetylcholine binding site and the high affinity site for noncompetitive blockers take place between topographically distinct sites and are, thus, indirect or allosteric.

V. CORRELATION OF ION CHANNEL OPENING WITH RAPID [³H]-CHLORPROMAZINE LABELING

Our approach to identify the structural elements involved in the ion channel gated by acetylcholine has been to compare the rapid kinetic properties of ion transport through the channel and [³H]-chlorpromazine labeling. First, the effect of chlorpromazine and phencyclidine, the compounds that we used in the structural studies, was investigated on single channel currents recorded by the patch-clamp method (Neher and Sakmann, 1976; Hamill et al., 1981) on C2 mouse myotubes (Changeux et al., 1986a). Under conditions where both compounds are positively charged (pH 7.2), 10–200 nM chlorpromazine or phen-cyclidine leads to shortened mean burst times as expected for high-affinity channel blockers (slow channel block, Neher and Steinbach, 1978). Yet, this shortening appeared voltage-independent (at variance with the effect of many, but not all, channel blockers) and, in addition, did not vary in a simple linear manner with concentration. Even though the electrophysiological data obtained by this method are accounted for by a variety of models, the effects observed at low concentration are consistent with the notion that, in this range of concentrations, chlorpromazine and phencyclidine block the ion channel by steric hindrance and that their high affinity site is located in (or near) the ion channel.

The subsequent analysis of the interaction of [^{3}H]-chlorpromazine with the AChR is based on the observation that purified membrane preparations from both *Electrophorus electricus* (Kasai and Changeux, 1971) and *Torpedo* (Hazelbauer and Changeux, 1974) preserve the regulation of ion channel opening by acetylcholine as monitored by the translocation of tracer cations (review Heidmann *et al.*, 1983a; Cash *et al.*, 1985). The development of methods for rapid measurement, *in parallel*, of agonist binding (Heidmann and Changeux, 1979; Boyd and Cohen, 1980) and ion fluxes (review Cash *et al.*, 1985) with the same membrane preparation (Heidmann *et al.*, 1983; Neubig and Cohen, 1980) leads to the conclusion that the binding of agonists triggers the opening of the ion channel in a relatively *high range of concentrations* [(45–80 μM acetylcholine or Dansyl-C6-choline (a fluorescent agonist) and 300–800 μM carbamylcholine] consistent with the known local concentrations of acetylcholine in the synaptic cleft during transmission (see Katz and Miledi, 1977).

The kinetics of covalent attachment of chlorpromazine to the AChR was resolved in a rapid mixing photolabeling apparatus (Heidmann and Changeux, 1984, 1986). Rapid addition of [^{3}H]-chlorpromazine to AChR-rich membrane fragments followed by brief (5 msec) UV irradiation of the mixture results in the simultaneous covalent labeling of all four chains of the AChR in a time-dependent manner. The rate of labeling increases (10^2–10^3 fold) under conditions of simultaneous addition of chlorpromazine and acetylcholine (or agonist) and varies linearly with chlorpromazine concentration in the explored concentration range (up to 10 μM) as expected for a diffusion-controlled bimolecular binding reaction with an *on* rate constant close to 10^7 M^{-1} sec^{-1}. It increases with agonist concentration with apparent dissociation constants close to 30 μM for acetylcholine and 400 μM for carbamylcholine (Heidmann and Changeux, 1984, 1986). It is blocked by snake venom α toxins and does not occur with competitive antagonists.

Moreover, under conditions where rapid desensitization takes place (see Chapter VI), the rates of chlorpromazine incorporation decline with the time course (half-life 6 sec^{-1} for 100 μM acetylcholine, maximum 15 sec^{-1}) and concentration dependence reported for the rapid desensitization of the ion flux response measured by chemical kinetics (Cash *et al.*, 1985). Thus, the enhanced rate of chlorpromazine incorporation elicited by acetylcholine binding to the AChR takes place at the level of a site which becomes accessible by diffusion when the channel opens. The selective labeling of helix II by [³H]-chlorpromazine (Giraudat *et al.*, 1986, 1987) was done in the presence of cholinergic agonists but under equilibrium conditions. It is *not* known yet if the same amino acids are labeled after rapid mixing. If this turns out to be the case, then the putative transmembrane helix II from each subunit might plausibly contribute to the walls of the ion channel. At the membrane level, the transport of ions through the AChR would then take place in an uncharged environment (Devillers-Thiéry *et al.*, 1983; Noda *et al.*, 1983; see Furois-Corbin and Pullman, 1986). Following this scheme, the charged groups distributed at both ends of the helix II segments could participate in the cationic selectivity of the channel.

Such an organization differs from those postulated by models which give to an amphipathic charged helix a crucial role in ion transport (Guy, 1984; Finer-Moore and Stroud, 1984, 1986; Mishina *et al.*, 1985; Ratnam *et al.*, 1986) and has recently received support from electrophysiological experiments carried out with *Xenopus* oocytes where functional AChR channels are expressed after injection of subunit mRNAs. Replacement of the δ subunit by various *Torpedo* × calf chimaeric δ subunit mRNAs suggests that helix II and the segment located between helix II and helix III is involved in determining the conductance difference noticed between *Torpedo* and calf channel at low divalent ion concentration (Imoto *et al.*, 1986).

VI. STRUCTURAL EVIDENCE FOR THE REGULATION OF AChR RESPONSE BY DESENSITIZATION

In addition to the fast all-or-none opening or activation of the ion channel, AChR mediates a "high order" regulation of the ionic response. Prolonged exposure to exogenous acetylcholine causes, within seconds or minutes, a reversible decline of the conductance response to cholinergic agonists (Katz and Thesleff, 1957). This desensitization phenomenon comprises two main kinetic processes: a fast one at the rate of 2–7 per second and a slow one at the rate of 0.01–0.1 per second (Feltz and Trautmann, 1980; Sakmann et al., 1980; Magleby and Pallotta, 1981).

Desensitization is still observed after purification of postsynaptic membrane fragments and even after purification of the AChR and reconstitution into lipid bilayers (review Popot et al., 1981; Montal et al., 1986). It is thus an intrinsic property of the AChR protein and does not require any obligatory covalent modification. The molecular transitions of the AChR engaged in desensitization have been resolved by a variety of rapid mixing techniques with fluorescent agonists (Heidmann and Changeux, 1979; Prinz and Maelicke, 1983) or radioactive ligands (Boyd and Cohen, 1980) and by following intrinsic (Grünhagen and Changeux, 1976; Bonner et al., 1976) or extrinsic (Grünhagen and Changeux, 1976; Grünhagen et al., 1977; Dunn et al., 1980) changes of fluorescence.

Rapid mixing studies with the fluorescent agonist Dansyl-C6-choline (Heidmann and Changeux, 1979) and [^{3}H]-acetylcholine (Boyd and Cohen, 1980) show that, in the absence of preincubation with agonist, 20% of the receptor spontaneously exists in a state of high affinity (3 nM for acetylcholine and 2 nM for Dansyl-C6-choline), the remaining 80% exhibiting a low affinity for agonists. Beyond the details of the analysis, which are often linked to a particular kinetic model (see Changeux et al., 1984; Hucho, 1986), two major conformational transitions have been resolved between these states: a "slow" one, with an apparent rate constant of about 0.01 per second and an "intermediate" one, with an apparent rate constant of 2 per second (Dunn et al., 1980) to 50 per second (Heidmann and Changeux, 1980). The state stabilized after equilibration binds acetylcholine and Dansyl-C6-choline with the same high affinity as the 20% of the population in the membrane present before mixing with the agonist.

An exceptional advantage of the in vitro preparations of functional AChR is that they make possible a comparison of the kinetics of both agonist binding and permeability response in the same time scale and with the same membrane preparation. Such parallel measurements show that the slow transition coincides with the slow desensitization phase of the ionic response (Neubig et al., 1982; Heidmann et al., 1983) and that the intermediate transition monitored with Dansyl-C6-choline fits with the fast desensitization (Heidmann et al., 1983a). In addition, they demonstrate the conjecture of Katz and Thesleff (1957) that the high affinity state for agonists of the AChR is a desensitized state of the AChR (the slow one) where the ion channel is shut. The data are accounted for by a minimal four-state model (Neubig and Cohen, 1980; Heidmann and Changeux, 1980) which has already been extensively discussed (see Changeux et al., 1984; Changeux and Heidmann, 1987) where A corresponds to the active state with the channel open, and I and D correspond to the rapidly and slowly desensitized states, respectively. These states are discrete and interconvertible. Their respective dissociation constants would be at least 50 to 100 μM (R), less than 1 μM (I), and 3 to 5 nM (D) for acetylcholine (Boyd and Cohen, 1980) and Dansyl-C6-choline (Heidmann and Changeux, 1979, 1980). In agreement with the "concerted" model of allosteric transition (Monod et al., 1965), several of these states [A (Jackson, 1985) and D (Heidmann and Changeux, 1979)] have been shown to occur spontaneously prior to ligand binding. Yet, this oversimplified scheme does not deal with the presence of two different acetylcholine binding sites per receptor oligomer nor with the occurrence of substates of the active conformation (see Changeux et al., 1986).

VII. REGULATION OF THE "EFFICACY" OF AChR RESPONSE BY ALLOSTERIC EFFECTORS ACTING ON DESENSITIZATION

Among other consequences, the model predicts that the "efficacy" of the receptor to respond to acetylcholine pulses might be regulated by effectors which differentially stabilize the I and D state or change the transition rates towards these states (Changeux et al., 1976; Heidmann and Changeux, 1979, 1982; Changeux and Heidmann, 1987). The simplest situation is that where the effector stabilizes the desensitized conformation of the AChR when it binds to the *acetylcholine site*. Under physiological conditions, acetylcholine may play this role (Magleby and Pallotta, 1981) resulting in a "homosynaptic" regulation of postsynaptic response efficacy. Interestingly, the dissociation constant for the desensitized state fits with the nonquantal leak concentration of acetylcholine measured at the level of the postsynaptic membrane in the presence of esterase inhibitor (Katz and Miledi, 1977). Accordingly, the non-quantal release of neurotransmitter may exert a regulatory role in pre-setting the RD equilibrium to a finite steady-state value.

Regulation of desensitization may also take place *via allosteric sites* distinct from the agonist binding site as long as their affinity differs in at least some of the multiple conformational states of the AChR. The best characterized allosteric sites are the high *and* low affinity sites for noncompetitive blockers. The high affinity site has been related to the ion channel (see Chapters IV and V), yet it might still be available for regulatory ligands in its *closed channel* conformation where, despite a reduced accessibility, its affinity might still be higher than in the active conformation. The low affinity sites possess a different pharmacological specificity (Heidmann *et al.*, 1978, 1983b) and are readily available to regulate desensitization.

At equilibrium, the noncompetitive blockers reversibly enhance (to different extents) the binding of cholinergic ligands to the AChR site, converting the shape of their binding curve from a sigmoid to a hyperbola (Cohen *et al.*, 1974) and conversely, cholinergic ligands potentiate their binding (Grünhagen and Changeux, 1976; Krodel *et al.*, 1979; Heidmann *et al.*, 1983b). Under conditions of rapid mixing they enhance the rate of fast desensitization of the electrophysiological (Magazanik and Vyskocil, 1975; Magleby and Pollotta, 1981) and ion flux response (references in Maleque *et al.*, 1983; Karpen *et al.*, 1982) and, in parallel, accelerate the intermediate and slow transitions (Grünhagen and Changeux, 1976; Weiland *et al.*, 1977; Heidmann and Changeux, 1979; Heidmann *et al.*, 1983b; Cohen and Boyd, 1979; Young and Sigman, 1981, 1983; Sine and Taylor, 1982) of the membrane-bound AChR.

The endogenous counterparts of the noncompetitive blockers are not known but physiologically significant effectors such as Ca^{2+} ions and electrical potential regulate these transitions. Ca^{2+} ions accelerate desensitization and/or stabilize the high affinity state (Magazanik and Vyskocil, 1970; Cohen et al., 1974; Heidmann and Changeux, 1979; Oswald, 1983), in particular when applied to the cytoplasmic side of the membrane (Miledi, 1980). Hyperpolarization also accelerates rapid desensitization while depolarization has the opposite effect (references in Takeyasu et al., 1983).

In a survey of the effects of peptides on the peripheral nicotinic receptor carried out in this laboratory, a polypeptide hormone of 49 amino acids produced by the thymus, thymopoietin (references in Audhya et al., 1981) was found to affect AChR desensitization. Its effect was studied both by patch clamp method on C2 mouse muscle cells in culture and by fluorescence spectroscopy with Dansyl-C6-choline and T. marmorata AChR-rich membranes (Revah et al., 1987). Previous work indicated that thymopoietin causes a decrease of muscle action potential recorded by electromyography in rat (Goldstein, 1974) and binds to Torpedo AChR with a high affinity (K_D 4×10^{-10} M) and in an α-bungarotoxin-sensitive manner (Venkatasubramanian et al., 1986). Thymopoietin, in addition, accelerates desensitization via an original mechanism: its effect on desensitization becomes manifest only in the presence of Ca^{2+} ions while its binding to the agonist site does not require Ca^{2+}. Analysis of the recordings with outside-out patches further indicates that Ca^{2+} ions exert their effect when present on the cytoplasmic face of the AChR. In other words, the enhanced desensitization of the AChR to acetylcholine is elicited by the conjunction of two distinct chemical signals, Ca^{2+} and thymopoietin, on both sides of the membrane.

An implication of thymopoietin in the pathology of Myasthenia gravis has been suggested in addition, or as alternative to, an auto immune reaction to the AChR (Goldstein, 1974).

Covalent modifications such as phosphorylation have been shown, in the past, to regulate the allosteric transition of classical regulatory enzymes such as phosphorylase b (review Krebs and Beavo, 1979) and it was suggested that such an effect might occur in the case of the acetylcholine receptor (Changeux et al., 1983) and of the ion channels (review Nestler and Greengard, 1984). Interestingly, recent evidence indicates that phosphorylation of the AChR accelerates desensitization of the ion flux response (Albuquerque *et al.*, 1986; Middleton *et al.*, 1986; Huganir *et al.*, 1986). Covalent modifications may thus contribute to the regulation of AChR response in a prolonged time scale by modulating its desensitization transitions in addition to, or as alternative of, the reversible binding of allosteric ligands.

The allosteric transitions of the AChR between a resting "activable" and desensitized "inactive" conformations thus appear as built-in molecular mechanisms available for the regulation of synapse efficacy at the postsynaptic level by extra- and intra-cellular communication signals.

VIII. MOLECULAR ARCHITECTURE OF THE POSTSYNAPTIC MEMBRANE

The AChR protein possesses in addition structural features which permit its highly localized and restricted distribution in the sarcolemmal membrane. The molecular mechanisms involved in this "topological regulation" at the cell level have been investigated in this laboratory by studying the properties and development of the postsynaptic membrane in *Torpedo* electric organ and vertebrate neuromuscular junction, primarily in the chick (review Changeux 1979, 1981; Changeux *et al.*, 1987).

In the adult fish electroplaque, the AChR is detected exclusively on the innervated face of the cell (Bourgeois *et al.*, 1971, 1978). In the innervated skeletal muscle, it is selectively localized at the endplate (Lee and Tseng, 1966; Barnard *et al.*, 1971; Salpeter and Loring, 1986). The absolute density of α toxin site per μm^2 of subsynaptic membrane reaches values as high as 50,000 (Bourgeois *et al.*, 1972, 1978) under the nerve endings in *E. electricus* electroplaque and falls to at least 100 times smaller values away from the synapses. Similar high densities have been reported at the vertebrate motor endplate (Barnard *et al.*, 1971; Miledi and Potter, 1971; review Salpeter and Loring, 1986), yet exclusively at the tip of the folds of the subsynaptic apparatus. These absolute values are such that, accounting for the known dimensions of the receptor protein, this subneural membrane domain consists of a monolayer of receptor molecules in close contact with each other. Voltage-dependent Na^+ channels are also present with a high surface density (Beam *et al.*, 1985; Angelides, 1986; Dreyfus *et al.*, 1986) at vertebrate endplate, but, for reasons of space limitation, should be distributed in distinct membrane domains such as the walls and bottom of the postsynaptic folds.

On the cleft side, the subneural area of the sarcolemmal membrane is coated by the basal lamina where the acetylcholinesterase heavy forms (review Massoulié and Bon, 1982), N-CAM (Reiger *et al.*, 1985; Covault and Sanes, 1985) and endplate specific antigens (Sanes and Chiu, 1983) are densely accumulated.

On the cytoplasmic side, filamentous material appears associated with the postsynaptic membrane of *Torpedo* electroplaque (Rosenbluth, 1975; Cartaud *et al.*, 1981; Heuser and Salpeter, 1979; Kordeli *et al.*, 1986) and of the neuromuscular junction (Couteaux, 1981; Hirokawa and Heuser, 1982), where a constellation of cytoskeletal proteins, actin, α actinin, talin, vinculin, and filamin, have been identified (Bloch and Hall, 1983; Sealock *et al.*, 1986). Underlying the endplate, the sarcoplasm makes a small "eminence" where mitochondria and muscle nuclei referred to by Ranvier (1875) as "fundamental nuclei" appear densely accumulated (Couteaux, 1978).

To test for the stability of this supramacromolecular edifice, *E. electricus* electric organ was denervated. Within a week all the nerve terminals disappeared, yet high density patches of AChR with the distribution, shape, and dimension of former subneural areas persisted up to 52 days (Bourgeois *et al.*, 1973, 1978). A similar persistence of subneural receptor patches was also reported with the neuromuscular junction (Frank *et al.*, 1975). The organization of the adult postsynaptic membrane thus appears remarkably stable.

IX. REGULATION OF AChR METABOLISM DURING SYNAPSE FORMATION

The sophisticated architecture of the postsynaptic membrane results from a long and complex series of molecular processes which include regulation of gene expression and multiple post transcriptional phenomena (review Salpeter and Loring, 1985; Changeux *et al.*, 1987).

Before the arrival of the exploratory motor axons, the fusion of myoblasts into myotubes is accompanied by an important increase of receptor number which becomes diffusely distributed all along the myotubes, exhibits significant lateral motion (Axelrod et al., 1976), undergoes rapid turnover (metabolic half-life 17–22 hr) (Chang and Huang, 1975; Berg and Hall, 1976; Devreotes and Fambrough, 1975) and possesses a channel mean open time of 3–10 msec (Katz and Miledi, 1972; Neher and Sakmann, 1976). In the adult muscle, the AChR is highly localized at the endplate, immobile (Rousselet et al., 1979, 1982), turns over slowly (half life of 10 days or more), displays a channel mean open time (τ) 3–5 times shorter than the embryonic receptor and possesses an intrinsic conductance (γ) significantly larger (review Salpeter and Loring, 1985). The exact values and the time at which these changes take place vary in different muscles and in different species (for instance, in the chick, τ and γ do not change significantly throughout development (Schuetze, 1980)). In addition, the number of receptor molecules incorporated in the subsynaptic membrane steadily increases during development, in particular after birth (Salpeter and Loring, 1985).

The initial increase of AChR number which coincides with the fusion of myoblasts into myotubes can be reproduced with cultured myoblasts (review Merlie, 1984). It corresponds to a de novo synthesis of receptor molecules as demonstrated by the incorporation of radioactive (Merlie et al., 1975, 1978) or heavy isotope (Devreotes and Fambrough, 1975; Devreotes et al., 1977) labeled amino acids into the receptor protein without significant change of its degradation rate (Merlie et al., 1976). Chronic injection of neuromuscular blocking agents in ovo does not significantly affect this initial onset of AChR biosynthesis (Burden, 1977a,b; Betz et al., 1980).

A second phase of the evolution of AChR metabolism corresponds to the decrease of total receptor content noticed, for instance, after day 15 in chick embryo breast muscle (Betz et al., 1977, 1980; Burden, 1977a, b). Such decline may result either from an enhanced degradation of the receptor protein (Stent, 1973) or from a repression of its biosynthesis. The metabolic degradation rate of the receptor protein throughout this evolution does not change (Betz et al.., 1977, 1980; Burden, 1977). A repression of receptor biosynthesis thus takes place. Parallel measurements of receptor surface density in non-junctional areas of the muscle fiber show that the decrease of total receptor content co-

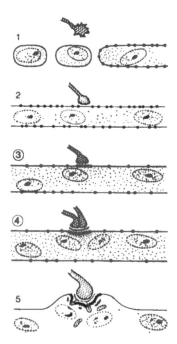

Fig. 2. Schematic representation of the evolution of the AChR in the course of the formation of the neuromuscular junction (black dots: AChR). 1) fusion of myoblasts into myotubes: the AChR biosynthesis is enhanced; 2) the exploratory motor axon approaches; 3) the growth cone contacts the myotube, a subneural cluster of AChR forms; 4) several motor nerve endings converge on the subneural cluster of AChR; 5) one motor nerve ending becomes stabilized; subneural folds develop; interactions with the cytoskeleton become apparent.

incides with the elimination of the AChR from these areas (Betz *et al.*, 1980).

At early stages of embryonic development, spontaneous movements appear (3.5 day in chick embryo) which are of neurogenic origin (Hamburger, 1970; Harris, 1981). Even growth cones release neurotransmitter and elicit an electrical response of the myotube membrane (review Chow and Poo, 1985). To test for the contribution of such neurally evoked electrical activity of the muscle fiber on the repression of receptor biosynthesis, chick embryos were chronically paralyzed by botulinum toxin (Giacobini-Robecchi *et al.*, 1975), *d*-tubocurarine (Burden, 1977a,b) or flaxedil (Bourgeois *et al.*, 1978; Betz *et al.*, 1980). Unambiguously, chronic paralyis maintains a high receptor content without changing its degradation rate. The neurally evoked activity of the embryonic muscle thus accounts for the repression of non-junctional receptor biosynthesis. A similar activity-dependent regulation can be demonstrated with myotubes in culture which exhibit a spontaneous (non-neurogenic) electrical activity (Shainberg and Burstein, 1976; Betz and Changeux, 1979), possibly as a result of an increased level of Ca^{2+}-regulated K^+ channel (Schmid-Antomarchi et al., 1985).

At the endplate level, chronic paralysis of chick embryos causes the disappearance of acetylcholinesterase (Giacobini et al., 1975; Betz et al., 1980) while the subsynaptic clusters of receptor persist (Betz et al., 1980; Lomo and Slater, 1980; review Massoulié and Bon, 1982; Lomo, 1987) suggesting a multifactorial regulation of the biogenesis of the postsynaptic domain by the nerve endings. Several attempts have been made to identify the neural factors involved by following the increase of AChR numbers on cultured muscle cells in the presence of extracts of nervous tissue and/or diverse substances (Podleski et al., 1978; Jessel et al., 1979; review Salpeter and Loring, 1985). It is not clear yet whether any of these factors are actually involved in endplate formation.

In this laboratory, we have studied peptides known to coexist with acetylcholine in spinal cord motor neurons (review Hökfelt et al., 1986). Calcitonin gene-related peptide (CGRP) is one of them (Takami et al., 1985; Fontaine et al., 1986; New and Mudge, 1986). Interestingly, application of CGRP (10^{-8} to 10^{-6} M) to cultured chick embryonic myotubes causes a 50% average increase in surface (and total) receptor number without significantly changing its degradation rate. This effect is not accompanied by a general increase in cellular protein biosynthesis, and is thus due to a specific enhancement of AChR biosynthesis (Fontaine et al., 1986; New and Mudge, 1986). It is independent of the effect produced by blocking the spontaneous electrical firing of the cultured myotubes by tetrodotoxin, but not of that caused by cholera toxin, which activates adenylate cyclase (Fontaine et al., 1986). These results suggest that CGRP and electrical activity regulate AChR numbers via different intracellular pathways.

The intracellular messengers which link electrical activity or CGRP binding to protein synthesis are not yet securely identified (review Changeux et $al.$, 1987; Salpeter and Loring, 1985). Ca^{2+}, which enters the voltage-sensitive Na^+ channel, has been suggested as a second messenger in the activity-dependent repression of AChR biosynthesis (Pezzementi and Schmidt, 1981; McManaman et $al.$, 1982; review Rubin, 1985) with the plausible relay of the diacylglycerol-phosphatidylinositol pathway (Berridge and Irvine, 1984). Depolarization, indeed, is known to induce turnover of inositol phospholipids in several tissues including smooth and skeletal muscle (references in Nishizuka, 1984; Ochs, 1986). The contribution of inositol phosphates and/or diacylglycerol in the electrical activity-dependent repression of extrajunctional AChR thus appears plausible but remains to be investigated.

A contribution of cyclic nucleotides has also been suggested in "membrane to gene" signalling. Addition of dibutyryl cAMP to chick muscle cells in culture increases the number of AChR (Betz and Changeux, 1979; Blosser and Appel, 1980), while dibutyryl cGMP has the opposite effect (Betz and Changeux, 1979; but McManaman et $al.$, 1982). A possible link between the Ca^{2+}-dependent and cyclic nucleotide-dependent regulatory pathways has been suggested (in the case of the regulation of Na^+ channels) (Sherman et $al.$, 1985). However, as mentioned above, the effects of cAMP derivatives and of adenylate cyclase activators are additive with those elicited by blocking electrical activity, suggesting a parallel rather than sequential effect of the respective second messengers (McManaman et $al.$, 1982; Fontaine et $al.$, 1986; Laufer and Changeux, 1987). Thus, while Ca^{2+} and/or inositol triphosphate and diacylglycerol may serve as second messengers for the activity-dependent repression of extra junctional AChR biosynthesis, cAMP might be the intracellular signal by which motoneuron-derived "anterograde" factors stimulate the accumulation of junctional receptor molecules. In support of this hypothesis, CGRP has been recently shown to stimulate accumulation of cAMP in skeletal muscle cells and to increase the rate of cAMP synthesis by muscle membranes. Moreover, elevation of cellular cAMP levels and of AChR number elicited by CGRP follow similar sustained time courses and occur in the same range of concentration (Laufer and Changeux, 1987). Yet, as in the case of other putative "anterograde" factors, we still ignorant CGRP is the physiological signal actually involved in the subneural accumulation of AChR. In this respect one may mention that positively charged latex beads elicit, in the absence of nerve, postsynaptic differentiations resembling those found in the mature motor endplates (Peng and Cheng, 1982). However, the possibility exists that binding of these "artificial" macromolecules to the surface of the myotube elicits the production of second messengers similar to those produced by more "physiological" neural factors.

X. ANALYSIS OF THE REGULATION OF AChR GENE EXPRESSION BY THE METHODS OF MOLECULAR GENETICS

The cloning and sequencing of cDNA and genes coding for the several subunits of the acetylcholine receptor from $Torpedo$ (Noda et al., 1983; Claudio et al., 1983; Devillers-Thiéry et al., 1983) and from other vertebrate species including humans (review Stroud and Finer-Moore, 1985) gave access to the analysis of AChR gene expression and regulation. In this laboratory, clones coding for the α subunit in chick have been isolated (Klarsfeld and Changeux, 1985) by cross-hybridization with $Torpedo$ probes (Devillers-Thiéry et al., 1983) on the basis of their high degree of sequence conservation (see also Ballivet et al., 1983). Southern blots of DNA digests from individual chicks disclosed restriction maps consistent with the presence of a *single copy gene* coding for the α subunit in the chick (Klarsfeld and Changeux, 1985).

Regulation of AChR gene expression may occur at several distinct levels such as transcription into mRNA, mRNA processing, mRNA stability, translation into polypeptide chains, assembly and processing of the subunits into the mature oligomer. Investigations carried out with C2 mouse myotubes (review Merlie and Smith, 1986) have shown that the initial burst of receptor biosynthesis which coincides with the fusion of myoblasts into myotubes is accompanied by an increase of the steady-state levels of AChR subunit mRNA which, in turn, is accounted for by an enhanced rate of transcription as determined by nuclear run-on experiments (Buonanno and Merlie, 1986).

The activity-dependent regulation of the α subunit gene expression which takes place in differentiated muscle fibers was analyzed in this laboratory in primary cultures of chick myotubes by Northern blot hybridization (Klarsfeld and Changeux, 1985). Blocking the spontaneous electrical activity by tetrodotoxin causes, after 1 and 2 days of application, a 4.5- and 13-fold increase respectively in α subunit mRNA levels while actin mRNA levels vary little (Klarsfeld and Changeux, 1985). These increases are of the same magnitude as those observed *in vivo* after denervation (see Merlie *et al.*, 1984; Goldman *et al.*, 1985 in the mouse). CGRP also elicits an increase of mRNA content but to a smaller extent (Fontaine *et al.*, 1987).

To approach the mechanisms involved in these regulations at the gene level, the 5'-end and promoter region of the α subunit gene was mapped and sequenced in the chick (Klarsfeld *et al.*, 1987). It includes a TATA and a CAAT box and a potential SP1 binding site. The mRNA start site was positioned by primer extension and S1 mapping experiments. The same site was found in both innervated and denervated muscle, a finding consistent with a control of AChR gene expression by muscle innervation at the level of transcription. When inserted in front of the chloramphenicol acetyltransferase gene, this promoter, including 850 bp of upstream sequence, directed high transient chloramphenicol acetyltransferase expression in transfected mouse C2.7 myotubes but not in C2.7 myoblasts or non myogenic 3T6 cells. Thus, the 850 bp-long, 5'-flanking region of the α subunit gene contains elements which confer tissue-specific and developmental control of expression. At this stage, however, it is not known whether or not this regulatory domain includes the sequences involved in the *cis*-regulation of the α subunit gene by electrical activity and CGRP.

Since the AChR is made up of four subunits, the expression of the four genes involved has to be coordinated (review Merlie and Smith, 1986) and this raises the possibility of a common control of their transcription by a common *cis*-regulatory sequence (Jacob and Monod, 1961). The linkage noticed between the chromosomic genes coding for the γ and δ subunits (Nef *et al.*, 1984) indeed pointed to such a possibility. To further investigate this question, the chromosomal localization of the *four* genes encoding the α, β, γ, and δ subunits was determined by a method recently developed by Robert *et al.* (1985), which is based upon the analysis of restriction fragment length polymorphisms between two mouse species, *Mus musculus domesticus* (DBA/2) and *Mus spretus* (MS). Analysis of the progeny of the interspecies mouse back cross (DBA/2 × SPE) × DBA/2 shows that the α-subunit gene cosegregrates with the α-cardiac actin gene on chromosome 17, that the β-subunit gene is located on chromosome 11 and that the γ and δ-subunit cosegregate and are located on chromosome 1 (Heidmann et al., 1986). Such partial dispersion of the subunit genes implies that their expression is regulated, at least in part, by trans-acting factors rather than by an "operon-like" mechanism.

Finally, a late enlargement (and complexification) of the post-synaptic membrane requires the maintenance of a significant biosynthesis of AChR in subneural areas after birth and in the adult. Expectedly, the levels of mRNA coding for the α and δ subunits have been found to be several times higher (2–14 fold) in synapse-rich than in synapse-free samples of mouse diaphragm, raising the possibility of a differential expression of receptor genes in junctional and non-junctional nuclei (Merlie and Sanes, 1985).

XI. MODEL OF AChR AND SYNAPTIC PROTEIN GENE REGULATION DURING THE DEVELOPMENT OF THE NEUROMUSCULAR JUNCTION

Many regulatory mechanisms control gene expression in the course of the terminal differentiation of skeletal muscle, yet only a few of them depend upon its functional innervation (review Changeux et al., 1987). On this basis, the proteins involved in skeletal muscle differentiation and motor endplate formation can be grouped into a minimum of three main families (Changeux et al., 1987).

A vast ensemble of proteins does not significantly change after denervation before the onset of muscle atrophy. This family I includes the "housekeeping proteins" and most (but not all) of the contractile proteins.

A second group of proteins referred to as family II evolves in a manner similar if not identical to the AChR and their "production" increases upon denervation. It includes particular species of the voltage sensitive Na^+ channel (Sherman and Catteral, 1984), and of the calcium sensitive K^+ channel (Schmid-Antomarchi et al., 1985), the cell adhesion molecule N-CAM (Rieger et al., 1985; Covault and Sanes, 1985; Covault et al., 1986), some basal lamina components (Sanes and Lawrence, 1983) and the neurite outgrowth promoting factor active on chick spinal neurons (MNGF) (Henderson et al., 1983). The cytoskeletal 43 KD protein ν_1 which selectively interacts with the AChR on its cytoplasmic face (see Chapter XII) is also a potential candidate for such a regulation which may also affect a few contractile proteins (Matsuda et al., 1984) and enzymes of energy metabolism (Lawrence and Salsgiver, 1983). It remains to be shown whether these postdenervation increases reflect a loosened electrical activity-dependent gene repression (see, however, Covault et al., 1986). Such *"negative"* regulation by electrical activity should of course be counteracted by "anterograde" factor(s) distinct from acetylcholine at the level of the endplate where these family II proteins persist in the adult.

A third group of proteins called *family III* is typified by the heavy 16–19.5S form of acetylcholinesterase which, in several species, is primarily localized at the endplate (Massoulié and Bon, 1982), disappears after chronic paralysis (Betz et al., 1980; Rubin et al., 1980; Vigny et al., 1976) *in vivo* or in co-cultures and reappears at this level after electrical stimulation (Lomo and Slater, 1980). In other words, an activity-dependent *"positive"* regulation takes place at the endplate level for this particular form of acetylcholinesterase (a distinct regulation may also take place for the other forms; see Massoulié and Bon, 1982).

The model we will discuss (Changeux et al., 1987) deals with the transcriptional regulation of genes coding for proteins from families II and III which have already been *determined*, i.e., where the chromatin is in a "ready to be transcribed" state. Such regulation involves a *minimum* of five distinct components: 1) "extracellular" first messengers, 2) "intracellular" second messengers, 3) *trans*-acting regulatory proteins binding to specific DNA regulatory sequences, 4) *cis*-acting DNA regulatory sequences, and 5) different categories of sarcoplasmic nuclei according to their topological distribution in subneural or in nonjunctional areas.

The proposal is that, within the same sarcoplasm, nuclei may exist under several distinct states of differentiation identified by the pattern of genes actually transcribed (and those switched off) which represent *"selections"* among the set of "open" or determined genes characteristic of the terminal state of skeletal muscle differentiation (see Changeux, 1986 for discussion).

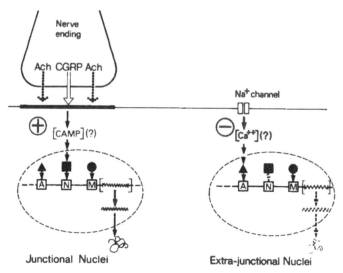

Fig. 3. Model of AChR gene regulation during the development of the neuromuscular junction. For explanation, see text.

In the *"myoblast"* nuclei, the transcription rates of the genes coding for families I, II, and III are negligible.

The *"embryonic"* myotube nuclei derive from the myoblast ones with the fusion of the myoblasts into myotubes. In this state (which may include several sub-states to account for the various patterns of contractile proteins synthesized) the genes coding for families I, II, and III are actively transcribed. The first and second messengers involved are not identified with certainty. A set of *trans*-acting proteins binding to homologous *cis*-acting sequences referred to as M (or muscle specific), should, at this stage, play a critical role.

In the *"adult extrajunctional nuclei,"* the genes of family I (and some of family III) are switched on, those of family II switched off. As a consequence, the muscle fiber becomes refractory to innervation. The first messenger is the electrical, rather than mechanical (Powell and Friedman, 1977) activity of the muscle fiber, the putative second messenger is Ca^{2+} (and possibly the diacylglycerol inositol-phosphates pathway), and the *trans*-acting proteins involved are assumed to bind putative *cis*-acting segments referred to as A.

In the *"adult junctional nuclei"*, the genes of family II are switched on as are those of family III (and possibly some of family I). The first messengers involved are expected to be multiple. Acetylcholine and/or electrical activity may activate the family III genes while those coding for the proteins of family II should be fully expressed. Coexisting neuronal messengers distinct from acetylcholine but released by the motor nerve endings such as CGRP (or ascorbate (Knaack and Podleski, 1985)) are expected to play a critical role. The second messengers involved are not known but cyclic AMP appears as one (among others) plausible candidate (Laufer and Changeux, 1987). Putative regulatory DNA sequences labeled N may be involved in this terminal differentiation of the sarcolemmal surface.

This model obviously corresponds to an oversimplified and schematized representation of the genetic regulatory mechanisms actually involved in motor endplate genesis and maturation. The gene families and states of the nuclei might be more numerous than postulated and might evolve in both sequential and parallel manners during development. For instance, if the change of mean channel open time, which occurs postnatally in amphibian and mammalian endplates results from a shift in the expression of the gene coding for the γ subunit to that coding for the ε subunit (Sakmann *et al.*, 1985; *but* Sakai *et al.*, 1985), then sub-states of the endplate nuclei have to be postulated between the newborn and the adult.

This simple-minded scheme raises, nevertheless, important questions. Since the genes of family II proteins are transcribed *both* in the non-innervated myotube by the embryonic nuclei and in the adult endplate by the subneural nuclei, are the *cis*-acting sequences involved identical or not in the two situations? In other words, are the M sequences distinct from the N ones in family II chromosomic genes?

Another original aspect of the model is the postulate that in the *same* cytoplasm nuclei may coexist with *different* states of differentiation. This situation imposes constraints on the production, diffusion, and degradation of the first and second messengers from and into the subneural region of the muscle fiber. It raises the question of the *reversibility* of the state of differentiation of the subneural nuclei, for instance, after denervation (see Loring and Salpeter, 1980; Salpeter and Loring, 1985) and, in a more general manner, that of the *stability* of the whole synapse. Family II includes components of the postsynaptic domain (AChR, 43 KD protein, . . .) but also proteins involved in cell surface adhesion (N-CAM, basal lamina antigens) and in transsynaptic "retrograde" signalling such as the MNGF(s). The maintenance of MNGF production by the endplate nuclei then may create a *positive* feedback loop upon the motor nerve ending which, in combination with an anterograde *positive* factor like CGRP, will constitute a closed circuit. The whole synapse will then be in a far-from equilibrium but stable steady-state which becomes resistant to protein turnover (for discussion see Changeux and Heidmann, 1987; Changeux *et al.*, 1987). Such a mechanism may be utilized for the selection of synapses in the course of development (see Gouzé *et al.*, 1983).

Several aspects of the model may be experimentally tested, for instance, by the identification of the first and second messengers involved in the regulation of the patterns of mRNA expressed in subneural *vs.* extra-junctional nuclei, by the analysis of the regional distribution of mRNA primary transcripts, for instance, by *in situ* hybridization, by the identification of the *cis*-acting DNA sequences and complementary *trans*-acting proteins, *etc.* . . The diverse *cis*-acting regulatory sequences involved might equally be located in 5'- or 3'-flanking regions as within introns of the chromosomic gene and could be identified, for instance, by DNase I mapping (Crowder and Merlie, 1986) and gene transfer experiments (Klarsfeld *et al.*, 1987). In this context, as already mentioned, some putative M sequences have been located on the α subunit gene in the 5' flanking regions within 850 bp (Klarsfeld *et al.*, 1987).

The extension of this model to nerve cells remains highly speculative (see Changeux, 1986). A major difference is, of course, the existence of a unique nucleus in neurons. Yet, different successive "states" of differentiation of the nucleus may exist from the committed neuroblast to the fully branched and connected adult neuron and the transitions from one state to the other may be regulated by a defined pattern of first messengers which include neurotransmitters, coexisting messengers (Hökfelt *et al.*, 1986), hormones and even the electrical firing of the neuron. An interesting issue becomes the existence or not of an analogy between the intracellular communication mechanisms involved in the membrane-to-gene coupling in muscle and in nerve cells and the eventual occurrence of "universal" mechanisms in activity-dependent regulation of gene expression.

XII. POST-TRANSLATIONAL REGULATION OF AChR TOPOLOGY

If transcriptional regulation of AChR gene expression contributes to the differentiation and maintenance of the postsynaptic domain of the motor endplate, it does not suffice to account for the highly anisotropic and stable distribution of the receptor molecules at the surface of the cell. Additional post-transcriptional phenomena have to be postulated which may include covalent modifications of the polypeptide chains, conformational maturation and assembly of the subunits, transport of the mature proteins to the cytoplasmic membrane, immobilization and anchoring at the subneural level, protection against rapid internalization and degradation (for review see Changeux, 1981; Merlie, 1983; Merlie and Smith, 1986; Changeux et al., 1986). Yet, many (if not all) components are already present in the developing myotube before innervation and the genesis of the postsynaptic membrane resembles an "assemblage" of pre-formed materials elicited by a few first and second messengers (see Changeux et al., 1981).

Here, the discussion will be limited to a protein discovered in this laboratory (Sobel et al., 1977, 1978), the 43 KD protein (review Changeux, 1981; Kordeli et al., 1986; Burden, 1985). This protein was first identified in AChR-rich membranes prepared from Torpedo electric organ (Sobel et al., 1977, 1978) as a major alkaline component on two-dimension gels (Saitoh and Changeux, 1981) in company with two other proteins of similar molecular weight identified respectively as cytosolic creatine phosphokinase (Barrantes et al., 1983; Gysin et al., 1983; Giraudat et al., 1984) and cytoplasmic actin (Gysin et al., 1983; Porter and Froehner, 1983). Selective proteolysis (Wennogle and Changeux, 1980), labeling with isotopic iodine (Saint John et al., 1982), and gold labeling with specific monoclonal antibodies (Nghiêm et al., 1983; Sealock et al., 1984) have shown that the 43 KD protein exclusively faces the cytoplasmic side of the membrane and distributes along with the AChR in approximate equimolar ratios (Sobel et al., 1978; La Rochelle and Froehner, 1986). Brief exposure of the receptor-rich membrane to pH 11 (Neubig et al., 1979) or lithium diiodo salicylate Elliott et al., 1980) releases the 43 KD protein and other peripheral proteins without significant changes of the receptor functional properties monitored by ion flux measurements or rapid binding of agonists and noncompetitive blockers (Neubig et al., 1979; Heidmann et al., 1980; Elliott et al., 1980). However, elimination of the 43 KD protein destabilizes the receptor to heat treatment (Saitoh et al., 1979) or proteolytic attack (Klymkowsky et al., 1980) and enhances its motion as monitored with spin-labeled (Rousselet et al., 1979, 1980, 1982) or phosphorescent (Lo et al., 1980) derivatives of α bungarotoxin and by electron microscopy (Barrantes et al., 1980; Cartaud et al., 1981). It binds to the cytoplasmic domain of the receptor, at least by way of the β subunit (Burden et al., 1983) and thus strongly immobilizes the molecule. A contribution of the heavy-form dimer of the AChR to this process has been suggested (Holton et al., 1984). A close association of the 43 KD protein with intermediate filaments containing desmin (Kordeli et al., 1986) leads to the suggestion that this highly insoluble, cystein-rich molecule (Sobel et al., 1978) may serve as an "intermediate piece" between the cytoskeleton and the postsynaptic membrane (Cartaud et al., 1983; Richardson and Witzemann, 1986; Kordeli et al., 1986).

The 43 KD protein can be phosphorylated in vitro (Saitoh and Changeux, 1980) and this opens the possibility of a regulation of the 43 KD protein attachment to the receptor and/or the cytoskeleton by phosphorylation-dephosphorylation reactions (Saitoh and Changeux, 1981; Gordon et al., 1983; Yéramian and Changeux, 1986).

A protein antigenically related to the 43 KD protein has been demonstrated in the postsynaptic membrane of mammalian muscle (Froehner et al., 1981; Froehner, 1984) and coaggregates with the AChR in cultures of Xenopus muscle cells (Burden, 1985; Peng and Froehner, 1985) or of rat myotubes (Bloch and Froehner, 1987) in the presence of motor innervation and even at the level of spontaneous patches.

The 43 KD protein may thus contribute to the immobilization, stabilization and change of functional and metabolic properties of the AChR which occur during the development and maturation of the postsynaptic membrane. Yet, many unknowns remain about its role in these processes.

XIII. CONCLUSIONS

Recent developments of the research on the functional organization of the AChR protein and on its contribution to the development of the postsynaptic domain of the electroplaque and motor endplate offer a variety of models for short- and long-term regulation of synapse properties at the postsynaptic level.

The acetylcholine receptor carries several categories of allosteric sites and undergoes a variety of conformational transitions. Similar features have been recently recognized with other ligand-regulated and/or voltage-sensitive ion channels such as the neuronal GABA-benzodiazepine receptor (Fischer and Olsen, 1986) or the voltage-sensitive Na^+ channel (Catterall, 1980). Depending on the initial balance between a resting R and an unresponsive desensitized D state and on the relative affinities of the ligand for these two states, allosteric effectors may lead to a "potentiation" or to a "depression" of the response. The effectors might be first messengers (neurotransmitters from neighboring synapses and/or coexisting neuronal messengers), intracellular second messengers and/or electrical potential. The integration of several of them in space and time *via* a "concerted" transition of the protein may naturally lead to plausible models of the Hebb synapse (Heidmann and Changeux, 1982; Changeux and Heidmann, 1986) and covalent modifications may extend these regulations to longer time scales (see Huganir *et al.*, 1986).

The activity-dependent regulation of AChR *number* observed in developing muscle may serve to model long-term changes of synapse properties in neurons. Yet, in nerve cells, the best evidence for activity-dependent regulation of gene expression concerns, up to now, the biosynthesis of neuropeptides or enzymes of neurotransmitter synthesis (review Black *et al.*, 1986). In addition, the topology of neurotransmitter receptors is far more complex than in muscle. Multiple species of receptors are synthesized by the same neuron but become segregated at the level of different categories of synapses. Such sophisticated topology requires post-transcriptional processes of "targeting" which may include a sorting out at the distal Golgi apparatus, a differential transport to the synapse and/or a local selection involving, for instance, homologs of the 43 KD protein and the cytoskeleton (see Changeux, 1986). Activity-dependent regulation of these processes may also take place at several different levels.

At all levels of such a "network" of molecular interactions, allosteric proteins integrate multiple regulatory signals and translate them into a biological response which varies from one level to the other. For *intercellular* signals, the proteins are membrane receptors and/or channel proteins and the short-term response a translocation of permeant ions or an enzyme activity which generates (or regulates) intracellular levels of second messengers. At the gene level, according to recent views (Ptashne, 1986), groups of DNA-binding proteins, interacting with each other in a cooperative manner *via* DNA loops, regulate the transcription of the adjacent genes. Such a macromolecular assembly may, as well, behave as an "allosteric" complex and integrate multiple regulatory signals, which would now be intracellular second messengers or the covalent modifications they trigger.

Finally, these complex regulations of neurotransmitter receptors, from their allosteric transitions to the multiple steps of their gene expression, may become resistant to the turnover of proteins and, thus, self-sustained (see Chapter XI). Several models for the perpetuation of activity-dependent changes of synapse properties have been proposed (see Crick, 1984; Changeux and Heidmann, 1987; Miller and Kennedy, 1986; Yéramian and Changeux, 1986). These highly hypothetical models may make less paradoxical the long-term storage of memories in the brain. Yet, their plausibility remains to be evaluated.

Acknowledgments

This work was supported by grants from the Muscular Dystrophy Association of America, the Fondation de France, the Collège de France, the Ministère de la Recherche, the Centre National de la Recherche Scientifique, the Institut National de la Santé et de la Recherche Médicale, and the Commissariat à l'Energie Atomique.

REFERENCES

Adams, P.R. (1981). Acetylcholine receptor kinetics. *J. Membr. Biol.* **58**, 161–174.
Albuquerque, E.X., Deshpande, S.S., Aracava, Y., Alkondon, M., and Daly, J.W. (1986). A possible involvement of cyclic AMP in the expression of desensitization of the nicotinic

acetylcholine receptor. *FEBS Lett.* **199**, 113–120.

Angelides, K.J. (1986). Fluorescently labeled Na+ channels are localized and immobilized to synapses of innervated muscle fibres. *Nature* **321**, 63–66.

Audhya, T., Schlesinger, D.H., and Goldstein, G. (1981). Complete amino acid sequences of bovine thymopoietins I, II, and III: closely homologous polypeptides. *Biochemistry* **20**, 6195–6200.

Axelrod, D., Ravdin, P.M., Koppel, D.E., Schlessinger, J., Webb, W.W., Elson, E.L., and Podleski, T.R. (1976). Lateral motion of fluorescently labeled acetylcholine receptors in membranes of developing muscle fibers. *Proc. Natl. Acad. Sci. U.S.A.* **73**, 4594–4595.

Ballivet, M., Patrick, J., Lee, J., and Heinemann, S. (1982). Molecular cloning of cDNA coding for the gamma-subunit of *Torpedo* acetylcholine receptor. *Proc. Natl. Acad. Sci. U.S.A* **79**, 4466–4470.

Ballivet, M., Nef, P., Stalder, R., and Fulpius, B. (1983). Genomic sequences encoding the alpha-subunit of acetylcholine receptor are conserved in evolution. *Cold Spring Harbor Symp. Quant. Biol.* **48**, 83–87.

Barnard, E.A., Wieckowski, J., and Chiu, T.H. (1971). Cholinergic receptor molecules and cholinesterase molecules at mouse skeletal muscle junctions. *Nature* **234**, 207–209.

Barrantes, F.J., Neugebauer, D.-Ch., and Zingsheim, H.P. (1980). Peptide extraction by alkaline treatment is accompanied by rearrangement of the membrane-bound acetylcholine receptor from *Torpedo marmorata*. *FEBS Lett.* **112**, 73–78.

Barrantes, F.J., Mieskes, G., and Wallimann, T. (1983). Creatine kinase activity in the *Torpedo* electrocyte and in the nonreceptor, peripheral nu-proteins from acetylcholine receptor-rich membranes. *Proc. Natl. Acad. Sci. U.S.A.* **80**, 5440–5444.

Beam, K.G., Caldwell, J.H., and Campbell, D.T. (1985). Na channels in skeletal muscle concentrated near the neuromuscular junction. *Nature* **313**, 588–590.

Berg, D.K. and Hall, Z.M. (1975). Loss of alpha-bungarotoxin from junctional and extra-junctional acetylcholine receptors in rat diaphragm muscle *in vivo* and in organ culture. *J. Physiol.* **252**, 771–789.

Berridge, M. and Irvine, R.F. (1987). Inositol triphosphate, a novel second messenger in cellular signal transactions. *Nature* **312**, 315–321.

Betz, H. and Changeux, J.P. (1979). Regulation of muscle acetylcholine receptor synthesis *in vitro* by derivatives of cyclic nucleotides. *Nature* **278**, 749–752.

Betz, H., Bourgeois, J.P., and Changeux, J.P. (1977). Evidence for degradation of the acetylcholine (nicotinic) receptor in skeletal muscle during the development of the chick embryo. *FEBS Lett.* **77**, 219–224.

Betz, H., Bourgeois, J.P., and Changeux, J.P. (1980). Evolution of cholinergic proteins in developing slow and fast skeletal muscle from chick embryo. *J. Physiol.* **302**, 197–218.

Black, I.B., Adler, J.E., and La Gamma, E.F. (1986). Impulse activity differentially regulates co-localized transmitters by altering messenger RNA level. *Prog. Brain Res.* **68**, 121–128.

Bloch, R.J. and Hall, Z.W. (1983). Cytoskeletal components of the vertebrate neuromuscular junction: vinculin, alpha actinin, and filamin. *J. Cell Biol.* **97**, 217–223.

Bloch, R.J. and Froehner, S.C. (1987). The relationship of the postsynaptic 43 K protein to acetylcholine receptors in receptor clusters isolated from cultured rat myotube. *J. Cell Biol.* (in press).

Blosser, J.C. and Appel, S.H. (1980). Regulation of acetylcholine receptor by cyclic AMP. *J. Biol. Chem.* **253**, 3088–3093.

Bon, F., Lebrun, E., Gomel, J., Van Rappenbusch, R., Cartaud, J., Popot, J.L., and

Changeux, J.P. (1984). Image analysis of the heavy form of the acetylcholine receptor from *Torpedo marmorata*. *J. Mol. Biol.* **176**, 205–237.

Bonner, R., Barrantes, F.J., and Jovin, T.M. (1976). Kinetics of agonist induced intrinsic fluorescence changes in membrane-bound acetylcholine receptor. *Nature* **263**, 429–431.

Boulter, J., Evans, K.L., Martin, G., Gardner, P.D., Connolly, J., Heinemann, S., and Patrick, J. (1986). Mouse muscle acetylcholine receptor molecular cloning of α, β, γ and δ-subunit cDNA's and expression in *Xenopus laevis* oocytes. *Abstr. Soc. Neurosci.* **12**, 40.2, 146.

Bourgeois, J.P., Tsuji, S., Boquet, P., Pillot, J., Ryter, A., and Changeux, J.P. (1971). Localization of the cholinergic receptor protein by immunofluorescence in eel electroplax. *FEBS Lett.* **16**, 92–94.

Bourgeois, J.P., Popot, J.L., Ryter, A., and Changeux, J.P. (1973). Consequences of denervation on the distribution of the cholinergic (nicotinic) receptor sites from *Electrophorus electricus* revealed by high resolution autoradiography. *Brain Res.* **62**, 557–563.

Bourgeois, J.P., Ryter, A., Menez, A., Fromageot, P., Boquet, P., and Changeux, J.P. (1972). Localization of the cholinergic receptor protein in *Electrophorus electroplax* by high resolution autoradiography. *FEBS Lett.* **25**, 127.

Bourgeois, J.P., Popot, J.L., Ryter, A., and Changeux, J.P. (1978). Quantitative studies on the localization of the cholinergic receptor protein in the normal and denervated electroplaque from *Electrophorus electricus*. *J. Cell Biol.* **79**, 200–216.

Boyd, N.D. and Cohen, J.B. (1980). Kinetics of binding of [^{3}H]acetylcholine and [^{3}H]carbamoylcholine to *Torpedo* postsynaptic membranes: slow conformational transitions of the cholinergic receptor. *Biochemistry* **19**, 5344–5358.

Brisson, A. and Unwin, P.N.T. (1985). Quaternary structure of the acetylcholine receptor. *Nature* **315**, 474–477.

Buonanno, A. and Merlie, J.P. (1986). Transcriptional regulations of nicotinic acetylcholine receptor genes during muscle development. *J. Biol. Chem.* **261**, 11452–11455.

Burden, S. (1977a). Development of the neuromuscular junction in the chick embryo. The number, distribution and stability of the acetylcholine receptor. *Dev. Biol.* **57**, 317–329.

Burden, S. (1977b). Acetylcholine receptors at the neuromuscular junction: Developmental change in receptor turnover. *Dev. Biol.* **61**, 79–85.

Burden, S.J., De Palma, R.L., and Gottesman, G.S. (1983). Crosslinking of proteins in acetylcholine receptor-rich membranes: Association between the β-subunit and the 43 Kd subsynaptic protein. *Cell* **35**, 687–692.

Burden, S.J. (1985). The subsynaptic 43 Kd protein is concentrated at developing nerve-muscle synapses *in vitro*. *Proc. Natl. Acad. Sci. U.S.A.* **82**, 7805–7809.

Cartaud, J., Benedetti, L., Cohen, J.B., Meunier, J.C., and Changeux, J.P. (1973). Presence of a lattice structure in membrane fragments rich in nicotinic receptor protein from the electric organ of *Torpedo marmorata*. *FEBS Lett.* **33**, 109–113.

Cartaud, J., Sobel, A., Rousselet, A., Devaux, P.F., and Changeux, J.P. (1981). Consequences of alkaline treatment for the ultrastructure of the acetylcholine-receptor-rich membranes from *Torpedo marmorata* electric organ. *J. Cell Biol.* **90**, 418–426.

Cartaud, J., Kordeli, C., Nghiêm, H.O., and Changeux, J.P. (1983). La proteine 43000 daltons: pièce intermédiaire assurant l'ancrage du recepteur cholinergique au cytosquelette sous-neural? *C. R. Acad. Sci. Paris* **297**, 285–289.

Cash, D., Aoshima, H., Pasquale, E., and Hess, G. (1985). Acetylcholine receptor-mediated ion fluxes in *Electrophorus electricus* and *Torpedo californica* membrane vesicles. *Rev. Physiol.*

Biochem. Pharmacol. **102**, 74–117.

Catterall, W.A. (1980). Neurotoxins that act on voltage-sensitive sodium channel. *Annu. Rev. Pharmacol. Toxicol.* **20**, 15–43.

Chang, C.C. and Huang, M.C. (1975). Turnover of junctional and extrajunctional acetylcholine receptors of the rat diaphragm. *Nature* **253**, 643–644.

Changeux, J.P. (1961). The feedback control mechanism of biosynthetic L-threonine deaminase by L-isoleucine. *Cold Spring Harbor Symp. Quant. Biol.* **26**, 313–318.

Changeux, J.P. (1966). Responses of acetylcholinesterase from *Torpedo marmorata* to salts and curarizing drugs. *Mol. Pharmacol.* **2**, 369–392.

Changeux, J.P. (1979). Remarks on the symmetry and cooperative properties of biological membranes. In *Symmetry and Function in Biological Systems at the Macromolecular Levels*, eds. Engström, A. and Strandberg, B. Nobel Symposium II, pp. 235–256. New York: John Wiley and Sons Inc.

Changeux, J.P. (1981). The acetylcholine receptor: An "allosteric" membrane protein. *Harvey Lect.* **75**, 85–254.

Changeux, J.P. (1986). Coexistence of neuronal messengers and molecular selection. *Prog. Brain Res.* **68**, 373–403.

Changeux, J.P. and Thiéry, J.P. (1968). On the excitability and cooperativity of biological membranes. In *Regulatory Functions of Biological Membranes*, ed. Jarnefelt, J., pp. 116-138. Amsterdam: Elsevier.

Changeux, J.P. and Heidmann, T. (1987). Allosteric receptors and molecular models of learning. In *Synaptic Function*, eds. Edelman, G., Gall, W.E., and Cowan, W.M. pp. 549–604. New York: Wiley.

Changeux, J.P., Podleski, T., and Wofsy, L. (1967a). Affinity labeling of the acetylcholine receptor. *Proc. Natl. Acad. Sci. U.S.A.* **58**, 2063–2070.

Changeux, J.P. Thiéry, J.P., Tung, Y., and Kittel, C. (1967b). On the cooperativity of biological membranes. *Proc. Natl. Acad. Sci. U.S.A.* **57**, 335–341.

Changeux, J.P., Benedetti, L., Bourgeois, J.P., Brisson, A., Cartaud, J., Devaux, P.F., Grünhagen, H.H., Moreau, M., Popot, J.L., Sobel, A., and Weber, M. (1976). Some structural properties of the cholinergic protein in its membrane environment relevant to its function as a pharmacological receptor. *Cold Spring Harbor Symp. Quant. Biol.* **40**, 203–210.

Changeux, J.P., Courrège, Ph., Danchin, A., and Lasry, J.M. (1981). Un mécanisme biochimique pour l'épigénèse de la jonction neuromusculaire. *C.R. Acad. Sci. Paris* **292**, 449–453.

Changeux, J.P., Bon, F., Cartaud, J., Devillers-Thiéry, A., Giraudat, J., Heidmann, T., Holton, B., Nghiêm, H.O., Popot, J.L., Van Rapenbusch, R., and Tzartos, S. (1983). Allosteric properties of the acetylcholine receptor protein from *Torpedo marmorata. Cold Spring Harbor Symp. Quant. Biol.* **48**, 35–52.

Changeux, J.P., Devillers-Thiéry, A., and Chemouilli, P. (1984). Acetylcholine receptor: an allosteric protein. *Science* **225**, 1335–1345.

Changeux, J.P., Giraudat, J., Dennis, M., Goeldner, M., Hirth, C., Mulle, C., Revah, F., Devillers-Thiéry, A., and Heidmann, T. (1987). Allosteric sites and conformational transitions of the acetylcholine receptor: models for short-term regulation of receptor response. In *Receptor-receptor Interactions: A New Intramembrane Integrative Mechanism*, eds. Fuxe, K. and Agnati, L. London: MacMillan Press (in press).

Changeux, J.P., Pinset, C., and Ribera, A.B. (1986a). Effects of chlorpromazine and phencyclidine on mouse C2 acetylcholine receptor kinetics. *J. Physiol.* **378**, 495–513.

Changeux, J.P., Klarsfeld, A., and Heidmann, T. (1987). The acetylcholine receptor and molecular models for short and long term learning. Dahlem Konferenzen. In *The Cellular and Molecular Bases of Learning*, eds. Changeux, J.P. and Konishi, M., pp. 31–83. London: Wiley.

Chow, I. and Poo, M.M. (1985), Release of acetylcholine from embryonic neurons upon contact with muscle cell. *J. Neurosci.* **5**, 1076–1082.

Claudio, T., Ballivet, M., Patrick, J., and Heinemann, S. (1983). Nucleotide and deduced amino acid sequences of *Torpedo californica* acetylcholine receptor gamma-subunit. *Proc. Natl. Acad. Sci. U.S.A.* **80**, 1111–1115.

Cohen, J.B. and Boyd, N.D. (1979). Conformational transitions of the membrane bound cholinergic receptor. In *Catalysis in Chemistry and Biochemistry*, eds. Pulman, B. and Ginsburg, O., pp. 293–304. D. Reidel Publ.

Cohen, J.B., Weber, M., and Changeux, J.P. (1974). Effects of local anesthetics and calcium on the interaction of cholinergic ligands with the nicotinic receptor protein from *Torpedo marmorata*. *Mol. Pharmacol.* **10**, 904–932.

Couteaux, R. (1978). Recherches morphologiques et cytochimiques sur l'organisation des tissus excitables, 225 pp. Paris: Robin et Marenge.

Couteaux, R. (1981). Structure of the subsynaptic sarcoplasm in the interfolds of the frog neuro-muscular junction. *J. Neurocytol.* **10**, 947–962.

Covault, J., Merlie, J., Goridis, C., and Sanes, J. (1986). Molecular forms of N-CAM and its RNA in developing and denervated skeletal muscle. *J. Cell Biol.* **102**, 731–739.

Criado, M., Hochschwender, S., Sarin, V., Fox, J.L., and Lindstrom, J. (1985). Evidence for unpredicted transmembrane domains in acetylcholine receptor subunits. *Proc. Natl. Acad. Sci. U.S.A.* **82**, 2004–2008.

Criado, M., Sarin, V., Fox, J.L., and Lindstrom, J. (1986). Evidence that the acetylcholine binding site is not formed by the sequence a 127–143 of the acetylcholine receptor. *Biochemistry* **25**, 2839–2846.

Crick, F. (1984). Memory and molecular turnover. *Nature* **312**, 101.

Crowder, C.M. and Merlie, J.P. (1986). DNase I-hypersensitive sites surround the mouse acetylcholine receptor δ-subunit gene. *Proc. Natl. Acad. Sci. U.S.A.* **83**, 8405–8409.

Dennis, M., Giraudat, J., Kotzyba-Hibert, F., Goeldner, M., Hirth, C., Chang, J.Y., and Changeux, J.P. (1986). A photoaffinity ligand of the acetylcholine binding site predominantly labels the region 179–207 of the alpha subunit on native acetylcholine receptor from *Torpedo marmorata*. *FEBS Lett.* **207**, 243–247.

Devillers-Thiéry, A., Giraudat, J., Bentaboulet, M., and Changeux, J.P. (1983). Complete mRNA coding sequence of the acetylcholine binding alpha subunit of *Torpedo marmorata* acetylcholine receptor: A model for the transmembrane organization of the polypeptide chain. *Proc. Natl. Acad. Sci. U.S.A.* **80**, 2067–2071.

Devreotes, P.N. and Fambrough, D.M. (1975), Acetylcholine receptor turnover in membranes of developing muscle fibers. *J. Cell Biol.* **65**, 335–358.

Devreotes, P.N., Gardner, J.M., and Fambrough, D.M. (1977). Kinetics of biosynthesis of acetylcholine receptor and subsequent incorporation into plasma membrane of cultured chick skeletal muscle. *Cell* **10**, 365–373.

Dreyfus, P., Rieger, F., Murawsky, M., Garcia, L., Loubet, A., Fosset, M., Pauron, D., Barhanin, J., and Lazdunski, M. (1986). The voltage-dependent sodium channel is colocalized with acetylcholine receptor at the vertebrate neuromuscular junction. *Biochem. Biophys. Res. Commun.* **139**, 196–201.

Dunn, S.M.J., Blanchard, S.G., and Raftery, M.A. (1980) Kinetics of carbamylcholine binding to membrane-bound acetylcholine receptor monitored by fluorescence changes of a covalently-bound probe. *Biochemistry* 19, 5652–5658.

Eldefrawi, A.T., Eldefrawi, M.E., Albuquerque, E.X., Oliveira, A.C., Mansour, N., Adler, M., Daly, J.W., Brown, G.B., Burgermeister, W., and Witkop, B. (1977). Perhydrohistrionicotoxin: a potential ligand for the ion conductance modulator of the acetylcholine receptor. *Proc. Natl. Acad. Sci. U.S.A.* 74, 2172–2176.

Elliott, J., Blanchard, S.G., Wu, W., Miller, J., Strader, C.D., Hartig, P., Moore, H.P., Racs, J., and Raftery, M.A. (1980). Purification of *Torpedo californica* post-synaptic membranes and fractionation of their constituent proteins. *Biochem. J.* 185, 667–677.

Feltz, A. and Trautmann, A. (1980). Interaction between nerve-released acetylcholine and bath applied agonists at the frog end plate. *J. Physiol.* 299, 533–552.

Finer-Moore, J. and Stroud, R.M. (1984). Amphipathic analysis and possible formation of the ion channel in an acetylcholine receptor. *Proc. Natl. Acad. Sci. U.S.A.* 81, 155–159.

Fischer, J.B. and Olsen, R.W. (1986). Biochemical aspects of GABA/Benzodiazepine receptor function. In *Benzodiazepine/GABA Receptors and Chloride Channels: Structural and Functional Properties*, pp. 241–259. New York: Alan R. Liss, Inc.

Fontaine, B., Klarsfeld, A., Hökfelt, T., and Changeux, J.P. (1986). Calcitonin gene-related peptide, a peptide present in spinal cord motoneurons, increases the number of acetylcholine receptors in primary cultures of chick embryo myotubes. *Neurosci. Lett.* 71, 59–65.

Frank, E., Gautvik, K., and Sommer-Schild, H. (1975). Cholinergic receptors at denervated mammalian endplates. *Acta Physiol. Scand.* 95, 66–76.

Froehner, S.C., Gulbrandsen, V., Hyman, C., Jeng, A.Y., Neubig, R.R., and Cohen, J.B. (1981). Immunofluorescence localization at the mammalian neuromuscular junction of the M_r 43,000 protein of *Torpedo* postsynaptic membranes. *Proc. Natl. Acad. Sci. U.S.A.* 78, 5230–5234.

Froehner, S.C. (1984). Peripheral proteins of postsynaptic membranes from *Torpedo* electric organ identified with monoclonal antibodies. *J. Cell Biol.* 99, 88–96.

Furois-Corbin, S. and Pullman, A. (1986). Theoretical studies of the packing of α-helices of poly (L-alanine) into transmembrane bundles. Possible significance for ion transfer. *Biochim. Biophys. Acta* 860, 165–177.

Giacobini-Robecchi, M.G., Giacobini, G., Filogamo, G., and Changeux, J.P. (1975). Effect of the type A toxin from *C. botulinum* on the development of skeletal muscles and on their innervation in chick embryo. *Brain Res.* 83, 107–121.

Giraudat, J., Devillers-Thiéry, A., Perriard, J.C., and Changeux, J.P. (1984). Complete nucleotide sequence of *Torpedo marmorata* mRNA coding for the 43,000 daltons nû₂ protein: muscle specific creatine kinase. *Proc. Natl. Acad. Sci. U.S.A.* 81, 7313–7317.

Giraudat, J., Dennis, M., Heidmann, T., Chang, J.Y., and Changeux, J.P. (1986). Structure of the high affinity binding site for noncompetitive blockers of the acetylcholine receptor: Serine-262 of the δ subunit is labeled by ^{3}H chlorpromazine. *Proc. Natl. Acad. Sci. U.S.A.* 83, 2719–2723.

Giraudat, J., Dennis, M., Heidmann, T., Haumont, P.T., Lederer, F., and Changeux, J.P. (1987). Structure of the high-affinity binding site for noncompetitive blockers of the acetylcholine receptor: [3]H chlorpromazine labels homologous residues in the beta and delta chains. *Biochemistry* 26, 2410–2418.

Giraudat, J., Devillers-Thiéry, A., Auffray, C., Rougeon, F., and Changeux, J.P. (1982). Identification of a cDNA clone coding for the acetylcholine binding subunit of *Torpedo marmorata* acetylcholine receptor. *EMBO J.* 1, 713–717.

Goeldner, M.P. and Hirth, C.G. (1980). Specific photoaffinity labeling induced by energy transfer: application to irreversible inhibition of acetylcholinesterase. *Proc. Natl. Acad. Sci. U.S.A.* **77**, 6439–6442.

Goldman, D., Boulter, J., Heinemann, S., and Patrick, J. (1985). Muscle denervation increases the levels of two mRNAs coding for the acetylcholine receptor alpha-subunit. *J. Neurosci.* **5**, 2553–2558.

Gordon, A.S., Milfay, D., and Diamond, I. (1983). Identification of a molecular weight 43,000 protein kinase in acetylcholine receptor-enriched membranes. *Proc. Natl. Acad. Sci. U.S.A.* **80**, 5862–5865.

Goldstein, G. (1974). Isolation of bovine thymin: a polypeptide hormone of the thymus. *Nature* **247**, 11–14.

Gouzé, J.L., Lasry, J.M., and Changeux, J.P. (1983). Selective stabilisation of muscle innervation during development: a mathematical model. *Biol. Cybern.* **46**, 207–215.

Grünhagen, H.H. and Changeux, J.P. (1976). Studies on the electrogenic action of acetylcholine with *Torpedo marmorata* electric organ. Quinacrine: a fluorescent probe for the conformational transitions of the cholinergic receptor protein in its membrane bound state. *J. Mol. Biol.* **106**, 497–516.

Grünhagen, H.H., Iwatsubo, M., and Changeux, J.P. (1977). Fast kinetic studies on the interaction of cholinergic agonists with the membrane-bound acetylcholine receptor from *Torpedo marmorata* as revealed by quinacrine fluorescence. *Eur. J. Biochem.* **80**, 225–242.

Guy, H.R. (1984). Structural model of the acetylcholine receptor channel based on partition energy and helix packing calculations. *Biophys. J.* **45**, 249–261.

Gysin, R., Yost, B., and Flanagan, S.D. (1983). Immunochemical and molecular differentiation of 43,000 molecular weight proteins associated with *Torpedo* neuroelectrocyte synapses. *Biochemistry* **22**, 5781–5789.

Hamburger, V. (1970) Embryonic mobility in vertebrates. In *The Neurosciences: Second Study Program*, ed. Schmidt, F.O., pp. 141–151. New York: Rockefeller Univ. Press.

Hamill, O.P., Marty, A., Neher, E., Sakmann, B., and Sigworth, F.J. (1981). Improved patch clamp techniques for high resolution current recording from cells and cell-free patches. *Pflügers Arch.* **391**, 85–100.

Haring, R., Kloog, Y., and Sokolovsky, M. (1984). Localization of phencyclidine binding sites on α and β subunits of the nicotinic acetylcholine receptor from *Torpedo ocellata* electric organ using azido phencyclidine. *J. Neurosci.* **4**, 627–637.

Harris, W.A. (1981). Neural activity and development. *Annu. Rev. Physiol.* **43**, 689–710.

Hazelbauer, G. and Changeux, J.P. (1974). Reconstitution of a chemically excitable membrane. *Proc. Natl. Acad. Sci. U.S.A.* **71**, 1479–1483.

Heidmann, O., Buonanno, A., Goeffroy, B., Robert, B., Guénet, J.L., Merlie, J.P., and Changeux, J.P. (1986). Chromosomal localization of the nicotinic acetylcholine receptor genes in the mouse. *Science* **234**, 866–868.

Heidmann, T. and Changeux, J.P. (1979). Fast kinetics studies on the interaction of a fluorescent agonist with the membrane-bound acetylcholine receptor from *Torpedo marmorata*. *Eur. J. Biochem.* **94**, 281–296.

Heidmann, T. and Changeux, J.P. (1980). Interaction of a fluorescent agonist with the membrane-bound acetylcholine receptor from *Torpedo marmorata* in the millisecond time range: resolution of an "intermediate" conformational transition and evidence for positive cooperative effects. *Biochem. Biophys. Res. Commun.* **97**, 889–896.

Heidmann, T. and Changeux, J.P. (1982). Un modèle moléculaire de régulation d'efficacité d'unsynapse chimique au niveau postsynaptique. *C. R. Acad. Sci. Paris* **295**, 665–670.

Heidmann, T., Sobel, A., Popot, J.L., and Changeux, J.P. (1980). Reconstitution of an acetylcholine receptor. Conservation of the conformational and allosteric transition and recovery of the permeability response; role of lipids. *Eur. J. Biochem.* **110**, 35–55.

Heidmann, T. and Changeux, J.P. (1984). Time-resolved photolabeling by the noncompetitive blocker chlorpromazine of the acetylcholine receptor in its transiently open and closed ion channel conformations. *Proc. Natl. Acad. Sci. U.S.A.* **81**, 1897–1901.

Heidmann, T. and Changeux, J.P. (1986). Characterization of the transient agonist-triggered state of the acetylcholine receptor rapidly labeled by the non-competitive blocker [^{3}H] chlorpromazine: additional evidence for the open channel conformation. *Biochemistry* **25**, 6109–6113.

Heidmann, T., Sobel, A., and Changeux, J.P. (1978). Recovery of allosteric interactions between a fluorescent cholinergic agonist and local anesthetics after removal of the detergent from cholate-solubilized membrane fragments rich in acetylcholine receptor. *FEBS Lett.* **94**, 397–404.

Heidmann, T., Bernhardt, J., Neumann, E., and Changeux, J.P. (1983a). Rapid kinetics of agonist binding and permeability response analysed in parallel on acetylcholine receptor-rich membranes from *Torpedo marmorata*. *Biochemistry* **22**, 5452–5459.

Heidmann, T., Oswald, R.E., and Changeux, J.P. (1983b). Multiple sites of action for non competitive blockers on acetylcholine receptor-rich membrane fragments from *Torpedo marmorata*. *Biochemistry* **22**, 3112–3127.

Henderson, C.E., Huchet, M., and Changeux, J.P. (1983). Denervation increases the neurite-promoting activity in extracts of skeletal muscle. *Nature* **302**, 609–611.

Heuser, J.E. and Salpeter, S.R. (1979). Organization of acetylcholine receptors in quick-frozen, deep-etched, and rotary-replicated *Torpedo* postsynaptic membrane. *J. Cell Biol.* **82**, 150–173.

Hirokawa, N. and Heuser, J.E. (1982). Internal and external differentiations of the postsynaptic membrane at the neuromuscular junction. *J. Neurocytol.* **11**, 487–510.

Hökfelt, T., Johansson, O., Ljungdahl, A., Lundberg, J.M., and Schultzberg, M. (1980). Peptidergic neurons. *Nature* **284**, 515–521.

Holton, B., Tzartos, S.J., and Changeux, J.P. (1984). Comparison of embryonic and adult *Torpedo* acetylcholine receptors by sedimentation characteristics and antigenicity. *Int. J. Dev. Neurosci.* **2**, 549–555.

Hucho, F.L., Oberthür, W., and Lottspeich, F. (1986). The ion channel of the nicotinic acetylcholine receptor is formed by the homologous helices M II of the receptor subunits. *FEBS Lett.* **205**, 137–142.

Hucho, F. (1986). The nicotinic acetylcholine receptor and its ion channel. *Eur. J. Biochem.* **158**, 211–226.

Huganir, R.L., Delcour, A.H., Greengard, P., and Hess, G.P. (1986). Phosphorylation of the nicotinic acetylcholine receptor regulates its rate of desensitization. *Nature* **321**, 744–776.

Hunkeler, W., Möhler, H., Pieri, L., Polc, P., Bonetti, E.P., Cumin, R., Schaffner, R., and Haefely, W. (1981). Selective antagonists of benzodiazepines. *Nature* **290**, 514–516.

Imoto, K., Methfessel, C., Sakmann, B., Mishina, M., Mori, Y., Konno, T., Fukuda, K., Kurasaki, M., Bujo, H., Fujita, Y., and Numa, S. (1986). Location of a δ-subunit region determining ion transport through the acetylcholine receptor channel. *Nature* **324**, 670–674.

Jackson, M.B. (1984). Spontaneous openings of the acetylcholine receptor channel. *Proc. Natl. Acad. Sci. U.S.A.* **81**, 3901–3904.

Jacob, F. and Monod, J. (1961). Genetic regulatory mechanisms in the synthesis of proteins. *J. Mol. Biol.* **3**, 318–356.

Jessel, T.M., Siegel, R.E., and Fischbach, G.D. (1979). Induction of acetylcholine receptors on cultured skeletal muscle by a factor extracted from brain and spinal cord. *Proc. Natl. Acad. Sci. U.S.A.* **76**, 5397–5401.

Kaldany, R. and Karlin, A. (1983). Reaction of quinacrine mustard with the acetylcholine receptor from *Torpedo californica*: Functional consequences and sites of labeling. *J. Biol. Chem.* **258**, 6232–6242.

Kao, P., Dwork, A., Kaldany, R., Silver, M., Wideman, J., Stein, S., and Karlin, A. (1984). Identification of the α-subunit half-cystine specifically labeled by an affinity reagent for the acetylcholine receptor binding site. *J. Biol. Chem.* **259**, 11662–11665.

Karlin, A. (1964). On the application of a plausible model of allosteric proteins to the receptor of acetylcholine. *J. Theor. Biol.* **16**, 306–320.

Karlin, A. (1980). Molecular properties of nicotinic acetylcholine receptors. In *Cell Surface Reviews*, eds. Poste, G., Nicolson, G.L., and Cotman, C.W., vol. 6, pp. 191–260. New York: Elsevier/North-Holland Biomed. Press.

Karlin, A. (1983). Anatomy of a receptor. *Neurosci. Comment.* **1**, 111–123.

Karlin, A. (1969). Chemical modification of the active site of the acetylcholine receptor. *J. Gen. Physiol.* **54**, 245–264.

Karpen, J.W., Aoshima, H., Abood, L.G., and Hess, G.P. (1982). Cocaine and phencyclidine inhibition of the acetylcholine receptor: Analysis of the mechanisms of action based on measurements of ion flux in the millisecond-to-minute time region. *Proc. Natl. Acad. Sci. U.S.A.* **79**, 2509–2513.

Kasai, M. and Changeux, J.P. (1971). *In vitro* excitation of purified membrane fragments by cholinergic agonists. I. Pharmacological properties of the excitable membrane fragments. II. The permeability change caused by cholinergic agonists. III. Comparison of the dose response curves to decamethonium with the corresponding binding curves of decamethonium to the cholinergic receptor. *J. Membr. Biol.* **6**, 1–80.

Katz, B. and Miledi, R. (1972). The statistical nature of the acetylcholine potential and its molecular components. *J. Physiol.* **224**, 665–699.

Katz, B. and Miledi, R. (1977). Transmitter leakage from motor nerve endings. *Proc. R. Soc. Lond.* **B 196**, 59–72.

Katz, B. and Thesleff, S. (1957). A study of the "desensitization" produced by acetylcholine at the motor end-plate. *J. Physiol.* **138**, 63–80.

Kistler, J. and Stroud, R.M. (1981). Crystalline arrays of membrane-bound acetylcholine receptor. *Proc. Natl. Acad. Sci. U.S.A.* **78**, 3678–3682.

Klarsfeld, A. and Changeux, J.P. (1985). Activity regulates the level of acetylcholine receptor alpha-subunit mRNA in cultured chick myotubes. *Proc. Natl. Acad. Sci. U.S.A.* **82**, 4558–4562.

Klarsfeld, A., Devillers-Thiéry, A., Giraudat, J., and Changeux, J.P. (1984). A single gene codes for the nicotinic acetylcholine receptor alpha-subunit in *Torpedo marmorata*: Structural and developmental implications. *EMBO J.* **3**, 35–41.

Klarsfeld, A., Daubas, P., Bourrachot, B., and Changeux, J.P. (1987). A 5′ flanking region of the chicken acetylcholine receptor alpha-subunit gene confers tissue-specificity and developmental control of expression in transfected cells. *Mol. Cell. Biol.* **7**, 951–955.

Klymkowsky, M.W., Heuser, J.E., and Stroud, R.M. (1980). Protease effects on the structure of acetylcholine receptor membranes from *Torpedo californica*. *J. Cell Biol.* **85**, 823–838.

Knaack, D. and Podleski, T. (1985). Ascorbic acid mediates acetylcholine receptor increase induced by brain extract on myogenic cells. *Proc. Natl. Acad. Sci. U.S.A.* **82**, 575–579.

Kordeli, E., Cartaud, J., Nghiêm, H.O., Pradel, L.A., Dubreuil, C., Paulin, D., and Changeux, J.P. (1986). Evidence for a polarity in the distribution of proteins from the cytoskeleton in *Torpedo marmorata* electrocytes. *J. Cell Biol.* **102**, 748–761.

Krause, K.L., Volz, K.W., and Lipscomb, W.N. (1985). Structure at 2.9-Å resolution of aspartate carbamoyltransferase complexed with the bisubstrate analogue N-(Phosphonacetyl)-L-aspartate. *Proc. Natl. Acad. Sci. U.S.A.* **82**, 1643–1647.

Krebs, G. and Beavo, J. (1979). Phosphorylation, dephosphorylation of enzymes. *Annu. Rev. Biochem.* **48**, 923.

Krodel, E.K., Beckman, R.A., and Cohen, J.B. (1979). Identification of a local anesthetic binding site in nicotinic post-synaptic membranes isolated from *Torpedo marmorata* electric tissue. *Mol. Pharmacol.* **15**, 294–312.

Kubo, T., Noda, M., Takai, T., Tanabe, T., Kayano, T., Shimizu, S., Tanaka, K., Takahashi, H., Hirose, T., Inayama, S., Kikuno, R., Miyata, T., and Numa, S. (1985). Primary structure of delta subunit precursor of calf muscle acetylcholine receptor deduced from cDNA sequence. *Eur. J. Biochem.* **149**, 5–13.

Langenbuch-Cachat, J., Bon, C., Mulle, C., Goeldner, M., Hirth, C., and Changeux, J.P. (1987). Photoaffinity labeling of the acetylcholine binding sites on the nicotinic receptor by an aryldiazonium derivative. Submitted.

La Rochelle, W.J. and Froehner, S.C. (1986). Determination of the tissue distributions and relative concentrations of the postsynaptic 43-kDa protein and the acetylcholine receptor in *Torpedo. J. Biol. Chem.* **261**, 5270–5274.

Laufer, R. and Changeux, J.P. (1987). Calcitonin gene-related peptide elevates cyclic AMP levels in chick skeletal muscle: possible neurotrophic role for a coexisting neuronal messenger. *EMBO J.* **6**, 901–906.

Lawrence, J.C. and Salsgiver, W.J. (1983). Levels of enzymes of energy metabolism are controlled by activity of cultured rat myotubes. *Am. J. Physiol.* **244**, C 348–355.

Lee, C.Y. and Tseng, L.F. (1966). Distribution of *Bungarus multicintus* venom following envenomation. *Toxicon* **3**, 281–290.

Lindstrom, J. (1986). Probing nicotinic acetylcholine receptors with monoclonal antibodies. *Trends Neurosci.* **9**, 401.

Lindstrom, J., Merlie, J.P., and Yogeeswaran, G. (1979). Biochemical properties of acetylcholine receptor subunits from *Torpedo californica. Biochemistry* **18**, 4465–4470.

Lo, M.M.S., Garland, P.B., Lamprecht, J., and Barnard, E.A. (1980). Rotational mobility of the membrane-bound acetylcholine receptor of *Torpedo* electric organ measured by phosphorescence depolarisation. *FEBS Lett.* **111**, 407–412.

Lomo, T. and Slater, C.R. (1980). Control of junctional acetylcholinesterase by neural and muscular influences in the rat. *J. Physiol.* **303**, 191–202.

Lomo, T. (1987). Formation of ectopic neuromuscular junctions: role of activity and other factors. Dahlem Konferenzen, *The Cellular and Molecular Bases of Learning*, eds. Changeux, J.P. and Konishi, M. pp. 359–373. London: Wiley.

Loring, R.H. and Salpeter, M.M. (1980). Denervation increases turnover rate of junctional acetylcholine receptors. *Proc. Natl. Acad. Sci. U.S.A.* **77**, 2293–2297.

McCarthy, M.P., Earnest, J.P., Young, E.F., Choe, S., and Stroud, R.M. (1986). The molecular neurobiology of the acetylcholine receptor. *Annu. Rev. Neurosci.* **9**, 383–413.

McCormick, D.J. and Atassi, Z. (1984). Localization and synthesis of the acetylcholine-binding site in the alpha-chain of the *Torpedo californica* acetylcholine receptor. *Biochem. J.* 224, 995-1000.

McManaman, J.L., Blosser, J.C., and Appel, S.H. (1982). Inhibitors of membrane depolarization regulate acetylcholine receptor synthesis by a calcium-dependent, cyclic nucleotide independent mechanism. *Biochim. Biophys. Acta* 720, 28-35.

Magazanik, L.G. and Vyskocil, F. (1970). Dependence of acetylcholine receptor desensitization on the membrane potential of frog muscle fibre and on the ionic changes in the medium. *J. Physiol.* 210, 507-518.

Magazanik, L.G. and Vyskocil, F. (1975). The effect of temperature on desensitization kinetics at the post-synaptic membrane of the frog muscle fibre. *J. Physiol.* 249, 285-300.

Magleby, K.L. and Pallotta, B.S. (1981). A study of desensitization of acetylcholine receptors using nerve-released transmitter in the frog. *J. Physiol.* 316, 225-250.

Maleque, M.A., Souccar, C., Cohen, J.B., and Albuquerque, E.X. (1982). Meproadifen reaction with the ionic channel of the acetylcholine receptor: potentiation of agonist-induced desensitization at the frog neuromuscular junction. *Mol. Pharmacol.* 22, 636-647.

Massoulié, J. and Bon, S. (1982). The molecular forms of cholinesterase and acetylcholinesterase in vertebrates. *Annu. Rev. Neurosci.* 5, 57-106.

Matsuda, R., Spector, D., and Strohman, R.C. (1984). Denervated skeletal muscle displays discoordinate regulation for the synthesis of several myofibrillar proteins. *Proc. Natl. Acad. Sci. U.S.A.* 81, 1122-1125.

Merlie, J. (1984). Biogenesis of the acetylcholine receptor, a multisubunit integral membrane protein. *Cell* 36, 573-575.

Merlie, J.P. and Smith, M.M. (1986). Synthesis and assembly of acetylcholine receptor, a multisubunit membrane glycoprotein. *J. Membr. Biol.* 91, 1-10.

Merlie, J.P., Isenberg, K.E., Russell, S.D., and Sanes, J.R. (1984). Denervation supersensitivity in skeletal muscle: Analysis with a cloned cDNA probe. *J. Cell Biol.* 99, 332-335.

Merlie, J.P., Sobel, A., Changeux, J.P., and Gros, F. (1975). Synthesis of acetylcholine receptor during differentiation of cultured embryonic muscle cells. *Proc. Natl. Acad. Sci. U.S.A.* 72, 4028-4032.

Merlie, J.P., Changeux, J.P., and Gros, F. (1978). Skeletal muscle acetylcholine receptor. Purification, characterization, and turnover in muscle cell cultures. *J. Biol. Chem.* 253, 2882-2891.

Merlie, J.P., Changeux, J.P., and Gros, F. (1976). Acetylcholine receptor degradation measured by pulse chase labeling. *Nature* 264, 74-76.

Merlie, J. and Sanes, J.R. (1985). Concentration of acetylcholine receptor mRNA in synaptic regions of adult muscle fibers. *Nature* 317, 66-68.

Middleton, P., Jaramillo, F., and Schuetze, M. (1986). Forskolin increases the rate of acetylcholine receptor desensitization at rat soleus endplates. *Proc. Natl. Acad. Sci. U.S.A.* 83, 4967-4971.

Miledi, R. (1980). Intracellular calcium and desensitization of acetylcholine receptors. *Proc. R. Soc. Lond.* B 209, 447-452.

Miledi, R. and Potter, L.T. (1971). Acetylcholine receptors in muscle fibers. *Nature* 233, 599-693.

Miller, S.G. and Kennedy, M.B. (1986). Regulation of brain type II Ca^{2+}/calmodulin-dependent protein kinase by autophosphorylation: A Ca^{2+}-triggered molecular switch. *Cell* 44, 861-870.

Mishina, M., Kurosaki, T., Tobimatsu, T., Morimoto, Y., Noda, M., Yamamoto, T., Terao,

M., Lindstrom, J., Takahashi, T., Kuno, M., and Numa, S. (1984). Expression of functional acetylcholine receptor from cloned cDNAs. *Nature* **307**, 604–608.

Mishina, M., Tobimatsu, T., Imoto, K., Tanaka, K., Fujita, Y., Fukuda, K., Kurasaki, M., Takahashi, H., Morimoto, Y., Hirose, T., Inayama, S., Takahashi, T., Kuno, M., and Numa, S. (1985). Location of functional regions of acetylcholine receptor alpha-subunit by site-directed mutagenesis. *Nature* **313**, 364–368.

Monod, J., Changeux, J.P., and Jacob, F. (1963). Allosteric proteins and cellular control systems. *J. Mol. Biol.* **6**, 306–328.

Monod, J., Wyman, J., and Changeux, J.P. (1965). On the nature of allosteric transitions: a plausible model. *J. Mol. Biol.* **12**, 88–118.

Montal, M., Anholt, R., and Labarca, P. (1986). The reconstituted receptor. In *Ion Channel Reconstitution*, ed. Miller, C., pp. 157–204. London: Plenum Press.

Muhn, P. and Hucho, F. (1983). Covalent labeling of the acetylcholine receptor from *Torpedo* electric tissue with the channel blocker [^{3}H]triphenylmethylphophonium by ultraviolet irradiation. *Biochemistry* **22**, 421–425.

Mulac-Jericevic, B. and Atassi, M.Z. (1986). Segment α-182-198 of *Torpedo californica* acetylcholine receptor contains a second toxin-binding region and binds anti-receptor antibodies. *FEBS Lett.* **199**, 68–74.

Nef, P., Mauron, A., Stalder, R., Alliod, C., and Ballivet, M. (1984). Structure, linkage and sequence of the two genes encoding the delta and gamma subunits of the nicotinic acetylcholine receptor. *Proc. Natl. Acad. Sci. U.S.A.* **81**, 7975–7979.

Neher, E. and Sakmann, B. (1976). Single channel currents recorded from membrane of denervated frog muscle fibers. *Nature* **260**, 799–802.

Neher, E. and Steinbach, J.H. (1978). Local anaesthetics transiently block currents through single acetylcholine-receptor channels. *J. Physiol.* **277**, 153–176.

Nestler, E. and Greengard, P. (1984). *Protein Phosphorylation in the Nervous System*. New York: Wiley.

Neubig, R.R. and Cohen, J.B. (1980). Permeability control by cholinergic receptors in *Torpedo* post synaptic membranes: Agonist dose response relations measured at second and millisecond times. *Biochemistry* **19**, 2770–2779.

Neubig, R.R., Boyd, N.D., and Cohen, J.B. (1982). Conformations of *Torpedo* acetylcholine receptor associated with ion transport and desensitization. *Biochemistry* **21**, 3460–3467.

Neubig, R.R., Krodel, E.K., Boyd, N.D., and Cohen, J.B. (1979). Acetylcholine and local anesthetic binding to *Torpedo* nicotinic postsynaptic membranes after removal of non-receptor peptides.. *Proc. Natl. Acad. Sci. U.S.A.* **76**, 690–694.

Neumann, D., Barchan, D., Safran, A., Gershoni, J.M., and Fuchs, S. (1986). Mapping of the α-bungarotoxin binding site within the α-subunit of the acetylcholine receptor. *Proc. Natl. Acad. Sci. U.S.A.* **83**, 3008–3011.

New, H.V. and Mudge, A.W. (1986). Calcitonin gene-related peptide regulates muscle acetylcholine receptor synthesis. *Nature* **323**, 809–811.

Nghiêm, H.O., Caraud, J., Dubreuil, C., Kordeli, C., Buttin, G., and Changeux, J.P. (1983). Production and characterization of a monoclonal antibody directed against the 43,000 M.W. nu$_1$ polypeptide from *Torpedo marmorata* electric organ. *Proc. Natl. Acad. Sci. U.S.A.* **80**, 6403–6407.

Nishizuka, Y. (1984). Turnover of inositol phospholipids and signal transduction. *Science* **225**, 1365–1370.

Noda, M., Takahashi, H., Tanabe, T., Toyosato, M., Furutani, Y., Hirose, T., Asai, M.,

Inayama, S., Miyata, T., and Numa, S. (1982). Primary structure of alpha-subunit precursor of *Torpedo californica* acetylcholine receptor deduced from cDNA sequence. *Nature* 299, 793-797.

Noda, M., Takahashi, H., Tanabe, T., Toyosato, M., Kikyotani, S., Hirose, T., Asai, M., Takashima, H., Inayama, S., Miyata, T., and Numa, S. (1983a). Primary structures of beta and delta-subunit precursors of *Torpedo californica* acetylcholine receptor deduced from cDNA sequences. *Nature* 301, 251-255.

Noda, M., Takahashi, H., Tanabe, T., Toyosato, M., Kikyotani, S., Furutani, Y., Hirose, T., Takashima, H., Inayama, S., Miyata, T., and Numa, S. (1983b). Structural homology of *Torpedo californica* acetylcholine receptor subunits. *Nature* 302, 528-532.

Noda, M., Furutani, Y., Takahashi, H., Toyosato, M., Tanabe, T., Shimizu, S., Kikyotani, S., Kayano, T., Hirose, T., Inayama, S., and Numa, S. (1983c). Cloning and sequence analysis of calf cDNA and human genomic DNA encoding α-subunit precursor of muscle acetylcholine receptor. *Nature* 305, 818-823.

Oberthür, W., Muhn, P., Baumann, H., Lottspeich, F., Wittmann-Liebold, B., and Hucho, F. (1986). The reaction site of a noncompetitive antagonist in the delta-subunit of the nicotinic acetylcholine receptor. *EMBO J.* 5, 1815-1819.

Ochs, R.S. (1986). Inositol trisphosphate and muscle. *Trends Biochem. Sci.* 1, 368-369.

Oswald, R.E. (1983). Effects of calcium on the binding of phencyclidine to acetylcholine receptor-rich membrane fragments from *Torpedo californica* electroplaque. *J. Neurochem.* 41, 1077.

Oswald, R., Sobel, A., Waksman, G., Roques, B., and Changeux, J.P. (1980). Selective labeling by [³H]-]trimethisoquin azide of polypeptide chains present in acetylcholine receptor rich membranes from *Torpedo marmorata. FEBS Lett.* 111, 29-34.

Oswald, R.E. and Changeux, J.P. (1981). Selective labeling of the delta-subunit of the acetylcholine receptor by a covalent local anesthetic. *Biochemistry* 20, 7166-7174.

Oswald, R. and Changeux, J.P. (1981). Ultraviolet light-induced labeling by noncompetitive blockers of the acetylcholine receptor from *Torpedo marmorata. Proc. Natl. Acad. Sci. U.S.A.* 78, 3925-3929.

Peng, H.B. and Cheng, P.C. (1982). Formation of postsynaptic specializations induced by latex beads in cultured muscle cells. *J. Neurosci.* 2, 1760-1774.

Peng, H.B. and Froehner, S.C. (1985). Association of the postsynaptic 43K protein with newly formed acetylcholine receptor clusters in cultured muscle cells. *J. Cell Biol.* 100, 1698-1705

Perutz, M.F., Fermi, G., Abraham, D.J., Poyart, C., and Bursaux, E. (1986). Hemoglobin as a receptor of drugs and peptides: X-ray studies of the stereochemistry of binding. *J. Am. Chem. Soc.* 108, 1064-1078.

Pezzementi, L. and Schmidt, J. (1981). Ryanodine alters the rate of acetylcholine receptor synthesis in chick skeletal muscle cell cultures. *J. Biol. Chem.* 256, 12.651-12.654.

Podleski, T.R., Axelrod, D., Ravdin, P., Greenberg, I., Johnson, M.M., and Salpeter, M.M. (1978). Nerve extract induces increase and redistribution of acetylcholine receptors on cloned muscle cells. *Proc. Natl. Acad. Sci. U.S.A.* 75, 2035-2039.

Popot, J.L., Cartaud, J., and Changeux, J.P. (1981). Reconstitution of a functional acetylcholine receptor: incorporation into artificial lipid vesicles and pharmacology of the agonist-controlled permeability changes. *Eur. J. Biochem.* 118, 203-214.

Porter, S. and Froehner, S.C. (1983). Characterization and localization of the $M_r = 43,000$ proteins associated with acetylcholine receptor-rich membranes. *J. Biol. Chem.* 258, 10034-

10040.

Powell, J.A. and Friedman, B.A. (1977). Electrical membrane activity: effect on distribution incorporation and degradation of acetylcholine receptors in the membranes of cultured muscle. *J. Cell Biol.* **75**, 321a.

Prinz, H. and Maelicke, A. (1983). Interaction of cholinergic ligands with the purified acetylcholine receptor protein: Equilibrium binding studies. *J. Biol. Chem.* **258**, 10263–10271.

Ptashne, M. (1986). Gene regulation by proteins acting nearby and at a distance. *Nature* **322**, 697–701.

Raftery, M.A., Hunkapiller, M., Strader, C.D., and Hood, L.E. (1980). Acetylcholine receptor: complex of homologous subunits. *Science* **208**, 1454–1457.

Ranvier, L. (1875). *Traité Technique d'Histologie.* Paris: Savy.

Ratnam, M., Le Nguyen, D., Rivier, J., Sargent, P.B., and Lindstrom, J. (1986). Transmembrane topography of nicotinic acetylcholine receptor: Immunochemical tests contradict theoretical predictions based on hydrophobicity profiles. *Biochemistry* **25**, 2633–2643.

Reynolds, J.A. and Karlin, A. (1978). Molecular weight in detergent solution of acetylcholine receptor from *Torpedo californica. Biochemistry* **17**, 2035–2038.

Revah, F., Mulle, C., Pinset, C., Audhya, T., Goldstein, G., and Changeux, J.P. (1987). Calcium-dependent effect of the thymic polypeptide thymopoietin on the desensitization of the nicotinic acetylcholine receptor. *Proc. Natl. Acad. Sci. U.S.A.* **84**, 3477–3481.

Richardson, G.P. and Witzemann, V. (1986). *Torpedo* electromotor system development: biochemical differentiation of *Torpedo* electrocytes *in vitro. Neuroscience* **17**, 1287–1296.

Rieger, F., Grumet, M., and Edelman, G. (1985). N-CAM at the vertebrate neuromuscular junction. *J. Cell Biol.* **101**, 285–293.

Robert, B., Barton, P., Minty, A., Daubas, P., Weydert, A., Bonhomme, F., Catalan, J., Chazottes, D., Guénet, J.L., and Buckingham, M. (1985). Investigation of genetic linkage between myosin and actin genes using an interspecific mouse back-cross. *Nature* **314**, 181–183.

Rosenbludt, J. (1975). Synaptic membrane structure in *Torpedo* electric organ. *J. Neurocytol.* **4**, 697–712.

Rousselet, A., Cartaud, J., and Devaux, P.F. (1979). Importance des interactions protéine-protéine dans le maintien de la structure des fragments excitables de l'organe électrique de *Torpedo marmorata. C.R. Acad. Sci. Paris* **D289**, 461–463.

Rousselet, A., Cartaud, J., Saitoh, T., Changeux, J.P., and Devaux, P. (1980). Factors influencing the rotational diffusion of the acetylcholine receptor-rich membranes from *Torpedo marmorata* investigated by saturation transfer electron spin resonance spectroscopy. *J. Cell Biol.* **90**, 418–426.

Rousselet, A., Cartaud, J., Devaux, P.F., and Changeux, J.P. (1982). The rotational diffusion of the acetylcholine receptor in *Torpedo marmorata* membrane fragments studied with a spin-labelled alpha-toxin: importance of the 43,000 protein(s). *EMBO J.* **1**, 439–445.

Rubin, L.L., Schuetze, S.M., Weill, C.L., and Fischbach, G.D. (1980). Regulation of acetylcholinesterase appearance at neuromuscular junctions *in vitro. Nature* **283**, 264–267.

Rubin, L.L. (1985). Increase in muscle Ca^{++} mediates changes in acetylcholinesterase and acetylcholine receptors caused by muscle contraction. *Proc. Natl. Acad. Sci. U.S.A.* **82**, 7121–7125.

Rubin, M.M. and Changeux, J.P. (1966). On the nature of allosteric transitions; implications of non exclusive ligand binding. *J. Mol. Biol.* **21**, 265–274.

Saint John, P.A., Froehner, S.C., Goodenough, D.A., and Cohen, J.B. (1982). Nicotinic postsynaptic membranes from *Torpedo*: Sidedness, permeability to macromolecules, and topography of major polypeptides. *J. Cell Biol.* **92**, 333–342.

Saitoh, T., Oswald, R., Wennogle, L.P., and Changeux, J.P. (1980). Conditions for the selective labelling of the 66,000 dalton chain of the acetylcholine receptor by the covalent non-competitive blocker 5-azido-³H- trimethisoquin. *FEBS Lett.* **116**, 30–36.

Saitoh, T. and Changeux, J.P. (1980). Phosphorylation *in vitro* of membrane fragments from *Torpedo marmorata* electric organ. *Eur. J. Biochem.* **105**, 51–62.

Saitoh, T. and Changeux, J.P. (1981). Change in state of phosphorylation of acetylcholine receptor during maturation of the electromotor synapse in *Torpedo marmorata* electric organ. *Proc. Natl. Acad. Sci. U.S.A.* **78**, 4430–4434.

Saitoh, T., Wennogle, L.P., and Changeux, J.P. (1979). Factors regulating the susceptibility of the acetylcholine receptor protein to heat inactivation. *FEBS Lett.* **108**, 489–494.

Sakmann, B., Patlak, J., and Neher, E. (1980). Single acetylcholine activated channels show burst-kinetics in presence of desensitizing concentrations of agonist. *Nature* **286**, 71–73.

Sakmann, B., Methfessel, C., Mishina, M., Takahashi, T., Takai, T., Kurasaki, M., Fukuda, K., and Numa, S. (1985). Role of acetylcholine receptor subunits in gating of the channel. *Nature* **318**, 538–543.

Salpeter, M. and Loring, R.H. (1985). Nicotinic acetylcholine receptors in vertebrate muscle: properties, distribution and neural control. *Prog. Neurobiol.* **25**, 297–325.

Sanes, J.R. and Lawrence, J.C., Jr. (1983). Activity-dependent accumulation of basal lamina by cultured rat myotubes. *Dev. Biol.* **97**, 123–136.

Schmid-Antomarchi, H., Renaud, J.F., Romey, G., Hughes, M., Schmid, A., and Lazdunski, M. (1985). The all-or-none role of innervation in expression of apamin receptor and of apamin sensitive Ca^{2+}-activated K^+ channel in mammalian skeletal muscle. *Proc. Natl. Acad. Sci. U.S.A.* **82**, 2188–2191.

Schramm, M. and Selinger, Z. (1984). Message transmission: receptor controlled adenylate cyclase system. *Science* **225**, 1350–1356.

Schuetze, E. (1980). The acetylcholine channel open time in chick muscle is not decreased following innervation. *J. Physiol.* **303**, 111–124.

Sealock, R., Paschal, B., Beckerle, M., and Burridge, K. (1986). Talin is a component of the mammalian neuromuscular junction. *Exp. Cell Res.* **163**, 143–150.

Sherman, S.J., Chrivia, J., and Catterall, W.A. (1985). Cyclic adenosine 3':5'-monophosphate and cytosolic calcium exert opposing effects on biosynthesis of tetrodotoxin—sensitive sodium channels in rat muscle cells. *J. Neurosci.* **5**, 1570–1576.

Sherman, S.J. and Catterall, W.A. (1984). Electrical activity and cytosolic calcium regulate levels of tetrodotoxin sensitive sodium channels in cultured rat muscle cells. *Proc. Natl. Acad. Sci. U.S.A.* **81**, 262–266.

Sine, S.M. and Taylor, P. (1982). Local anesthetics and histrionicotoxin are allosteric inhibitors of the acetylcholine receptor. Studies of clonal muscle cells. *J. Biol. Chem.* **257**, 8106–8114.

Sobel, A., Weber, M., and Changeux, J.P. (1977). Large scale purification of the acetylcholine receptor protein in its membrane-bound and detergent extracted forms from *Torpedo marmorata* electric organ. *Eur. J. Biochem.* **80**, 215–224.

Sobel, A., Heidmann, T., Hofler, J., and Changeux, J.P. (1978). Distinct protein components from *Torpedo marmorata* membranes carry the acetylcholine receptor site and the binding site for local anesthetics and histrionicotoxin. *Proc. Natl. Acad. Sci. U.S.A.* **75**, 510–514.

Stent, G. (1973). A physiological mechanism for Hebb's postulate of learning. *Proc. Natl.*

Acad. Sci. U.S.A. **70**, 997–1001.

Stroud, R.M. and Moore, J.F. (1985). Acetylcholine receptor structure, function, and evolution. *Annu. Rev. Cell Biol.* **1**, 317–351.

Sumikawa, K., Houghton, M., Smith, J.C., Bell, L., Richards, B.M., and Barnard, E.A. (1982). The molecular cloning and characterization of cDNA coding for the alpha subunit of the acetylcholine receptor. *Nucleic Acids Res.* **10**, 5809–5822.

Takai, T., Noda, M., Mishina, M., Shimizu, S., Furutani, Y., Kayano, T.T., Ikeda, T., Kubo, T., Takahashi, H., Takahashi, T., Juno, M., and Numa, S. (1985). Cloning, sequencing and expression of cDNA for a novel subunit of acetylcholine receptor from calf muscle. *Nature* **315**, 761–764.

Takami, K., Kawai, Y., Uchida, S., Tokuyama, M., Shiotani, Y., Yoshida, H., Emson, P.C., Girgis, S.H., Hillyard, C.J., and MacIntyre, I. (1985). Effect of calcitonin gene-related peptide on contraction of striated muscle in the mouse. *Neurosci. Lett.* **60**, 227–230.

Takeyasu, K., Udgaonkar, J.B., and Hess, G.P. (1983). Acetylcholine receptor: Evidence for a voltage-dependent regulatory site for acetylcholine. Chemical kinetic measurements in membrane vesicles using a voltage clamp. *Biochemistry* **22**, 5973–5978.

Venkatasubramanian, K., Audhya, T., and Goldstein, G. (1986). Binding of thymopoietin to acetylcholine receptor. *Proc. Natl. Acad. Sci. U.S.A.* **83**, 3171–3174.

Vigny, M., Digiamberardino, L., Couraud, J.Y., Rieger, F., and Koenig, J. (1976). Molecular forms of chicken acetylcholinesterase: effect of denervation. *FEBS Lett.* **69**, 277–280.

Weber, M. and Changeux, J.P. (1974a). Binding of *Naja nigricollis* ^{3}H-alpha-toxin to membrane fragments from *Electrophorus* and *Torpedo* electric organs. 1. Binding of the tritiated alpha-neurotoxin in the absence of effector. *Mol. Pharmacol.* **10**, 1–14.

Weber, M. and Changeux, J.P. (1974b). Binding of *Naja nigricollis* ^{3}H-alpha-toxin to membrane fragments from *Electrophorus* and *Torpedo* electric organs. 3. Effect of the cholinergic agonists and antagonists on the binding of the tritiated α-neurotoxin. *Mol. Pharmacol.* **10**, 15–34.

Weber, M. and Changeux, J.P. (1974c). Binding of *Naja nigricollis* ^{3}H-alpha-toxin to membrane fragments from *Electrophorus* and *Torpedo* electric organs. 3. Effects of local anaesthetics on the binding of the tritiated α-neurotoxin. *Mol. Pharmacol.* **10**, 35–40.

Weiland, G., Frisman, D., and Taylor, P. (1979). Affinity labeling of the subunits of the membrane associated cholinergic receptor. *Mol. Pharmacol.* **15**, 213–226.

Weiland, G., Georgia, B., Lappi, S., Chignell, C.F., and Taylor, P. (1977). Kinetics of agonist-mediated transitions in state of the cholinergic receptor. *J. Biol. Chem.* **252**, 7648–7656.

Wennogle, L.P. and Changeux, J.P. (1980). Transmembrane orientation of proteins present in acetylcholine receptor-rich membranes from *Torpedo marmorata* studied by selective proteolysis. *Eur. J. Biochem.* **106**, 381–393.

Wilson, P.T., Lentz, T.L., and Hawrot, E. (1985). Determination of the primary amino acid sequence specifying the alpha-bungarotoxin binding site on the alpha subunit of the acetylcholine receptor from *Torpedo californica*. *Proc. Natl. Acad. Sci. U.S.A.* **82**, 8790–8794.

Wise, D.S., Schoenborn, B.P., and Karlin, A. (1981). Structure of acetylcholine receptor dimer determined by neutron scattering and electron microscopy. *J. Biol. Chem.* **256**, 4124–4126.

Yéramian, E. and Changeux, J.P. (1986). Un modèle de changement d'efficacité synaptique à long terme fondé sur l'interaction du récepteur de l'acétylcholine avec la protéine sous synaptique de 43,000 daltons. *C.R. Acad. Sci. Paris* **302**, 609–616.

Young, A.P. and Sigman, D.S. (1981). Allosteric effects of volatile anesthetics on the membrane-bound acetylcholine receptor protein. I. Stabilization of the high-affinity state. *Mol. Pharmacol.* **20**, 498–505.

Young, A.P. and Sigman, D.S. (1983). Conformational effects of volatile anesthetics on the membrane-bound acetylcholine receptor protein: facilitation of the agonist-induced affinity conversion. *Biochemistry* **22**, 2154–2161.

3

NERVE GROWTH FACTOR: ITS ROLE IN NEURONAL GROWTH AND MAINTENANCE

ERIC M. SHOOTER, MONTE J. RADEKE, THOMAS P. MISKO, AND STUART C. FEINSTEIN

Department of Neurobiology, Stanford University School of Medicine, Stanford, CA 94305, U.S.A.

The discovery of nerve growth factor (NGF) moved developmental neurobiology into the molecular era and allowed a number of phenomenological observations to be explained in molecular terms. It was known prior to the discovery of NGF that the process of neuronal cell death determines the final size of a given neuronal population and that cell death occurs at the time that the nerve fibers reach their targets (*16*). Furthermore, the size of the target influences the extent of cell death suggesting that the targets secrete trophic factors which neurons require for survival. NGF is the first of these trophic factors to be identified and purified. The initial experiments which led to this discovery involved transplantation of mouse sarcomas to the body wall of chick embryos and the observation that the sympathetic ganglia became hypertrophied and vigorously sprouted neurites (*27*). Since grafting of the sarcoma onto the chorioallantoic membrane gave the same results it was inferred that the sarcoma secreted a diffusible factor (*28*). In tissue culture it was observed that extracts of the sarcoma enhanced the survival of sensory and sympathetic ganglia and stimulated neurite outgrowth (29). This in turn provided the basis of a biological assay for the purification of the active factor (NGF), and the preparation of NGF antibody (6). The critical *in vivo* dependency of sympathetic neurons on NGF was then demonstrated by the failure of virtually all sympathetic neurons to develop when NGF antibody was injected into newborn mice, rats or other mammals (26). More recent experiments also showed that exposure of fetal rats to NGF antibody *via* placental transfer significantly reduced the number of sensory neurons which developed normally (20).

Research over the past two decades has amply confirmed that NGF is a trophic factor, that it is synthesized in the targets of the NGF-responsive neurons and that it determines the selective survival and maintenance of the neurons which innervate these targets. The recent finding that NGF synthesis in the target fields of trigeminal sensory neurons only begins when the axons reach these targets (7, 22) fits well into this scheme. Moreover, since the level of NGF synthesis in the target is very low it is insufficient to support all the ingrowing fibers and so the process of selective regulation of survival occurs (18, 37).

I. THE FLOW OF NGF FROM TARGET TO CELL BODY

A flow of NGF from the targets to the cell bodies of the NGF-responsive neurons has been defined (Fig. 1). The NGF, synthesized in the target organs is secreted, sequestered by the nerve terminals, transported retrogradely and delivered as intact, biologically active NGF within vesicles, to the cell bodies of mature sympathetic and sensory neurons (42). Interruption of this flow by removal of the target organ, damage to the nerve terminal or axon or disruption of the axonal microtubules (24) results in degeneration of the neurons, a process which is reversed by injecting an excess of NGF at the time of injury (17, 42). Thus the continuous movement of NGF from the target to the cell body (23) is the basis of NGF's ability, *in vivo*, to determine neuronal survival, regulate axon growth or regrowth and maintain the differentiated phenotype. The retrograde flow of NGF is initiated by its interaction with specific NGF receptors on the nerve terminals. The receptors and their bound NGF are internalized in the process of receptor med-iated endocytosis and the NGF containing vesicles transported along

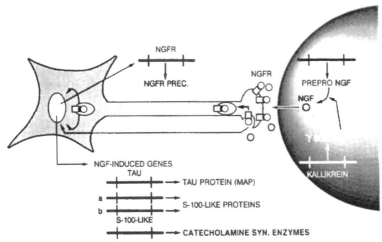

Fig. 1. The NGF pathway. The figure summarizes the background of the physiological flow of NGF from the target (on the right) to the cell body of NGF responsive neurons. The biosynthesis of NGF in the target is assumed to follow the pattern found in the mouse submaxillary gland (a sympathetically innervated target). Only one NGF receptor is shown at the nerve terminals although, as described in the text, two classes of receptors are present and are involved in the binding and internalization of NGF. The flow of NGF and NGF receptor-containing vesicles is along microtubules. The black arrows pointing to the nucleus indicate possible sites of intracellular signal generation: the receptor at the nerve terminal, the receptor in the vesicle or a second messenger generated at the nerve terminal receptor that affect, finally, gene transcription. Several of the genes whose transcription is induced by NGF are shown under the neuron. The regulation of the transcription of the NGF receptor gene has not yet been defined, nor has that of the NGF gene in the target.

microtubules to the cell body. Here the vesicles fuse finally with lysosomal structures and the NGF is quite rapidly degraded (24). The intracellular signals which mediate the function of NGF are generated somewhere in this pathway. Since a unique feature of NGF's action is its internalization in the periphery and its retrograde transport, it is possible that signals arise during this process as well as by transduction at the level of the receptor.

II. THE NGF RECEPTORS AND SIGNAL TRANSDUCTION

Since its initial purification (48) much has been learnt about the biochemistry of NGF. The active NGF molecule is a basic protein, a dimer of two identical chains, generated from a precursor coded for by a single gene (*36, 43*). In the mouse submaxillary gland one of the proteins associated with NGF, the serine protease γ subunit, is a processing enzyme for the precursor.

Besides primary sensory and sympathetic neurons a useful model system for investigations into the molecular basis of the actions of NGF is the rat pheochromocytoma cell line PC12 (*11*). Whereas PC12 cells grow and divide in an appropriate medium, addition of NGF results in gradual cessation of cell division and the acquisition by the PC12 cells of a number of characteristics of sympathetic neurons including neurite outgrowth. Two types of receptors are found on primary sensory and sympathetic neurons and PC12 cells (*47*) (Table I). They differ in their affinities for NGF (approx. 10^{-11} and 10^{-9} M, respectively), their numbers (approx. 10 times as many of the lower compared to the higher affinity receptor) and their dissociation kinetics (dissociation from the lower affinity receptor is approximately 100-fold faster than from the higher affinity receptor). In addition, NGF-occupied high affinity or slow receptors (using the terminology suggested in ref. *35* based on dissociation rates) are trypsin resistant and insoluble in Triton X-100 while occupied low affinity or fast receptors are trypsin labile and Triton X-100 soluble. The rate of dissociation of ^{125}I-NGF from the slow NGF receptors is enhanced by NGF under conditions where receptor occupancy is increased or decreased (*40*). The above data can be explained by the mobile receptor hypothesis by assuming that the fast NGF receptors interact

TABLE I

The Properties of the Two Classes of NGF Receptors

Property	S-NGFR (H-NGFR)	F-NGFR (L-NGFR)
$K_d(^{125}\text{I-NGF binding})$	10^{-11} M	10^{-9} M
Dissociation $(t_1/_2)$	$\sim$10 min	$\sim$3 sec
(^{125}I-NGF release)	Slow	Fast
Trypsin (occupied receptor)	Stable	Labile
Triton X-100 (occupied receptor)	Insoluble	Soluble
M_r MW complex	158 K	100 K
receptor	140 K	80 K

reversibly with effector molecules in the cell membrane to alter receptor affinity and other properties (*e.g.*, trypsin sensitivity). Such a model assumes that NGF induces a conformational (or other) changes in the receptor which enables the NGF-fast NGF receptor complex to recognize the effector protein. Evidence for a ligand-induced conversion of the fast to the slow NGF receptor has been presented (*25*).

Conversion of the fast NGF receptor to a slow receptor form (not necessarily identical with the naturally occurring slow receptor) occurs when crosslinking agents such as wheat germ agglutinin (WGA) or anti-NGF antibody are added after the binding of NGF to PC12 cells or sensory neurons (*4, 5, 12, 45, 46*). This conversion not only produces trypsin-resistance in the receptor but it also renders it insoluble in Triton X-100, raising the possibility that clustering of the NGF receptors involves attachment directly or indirectly (*44*) to the cytoskeleton, implying that the effector molecule may be a cytoskeleton and membrane attached protein.

What is the functional significance of the two receptor types? Dose response curves for neurite outgrowth from all responsive cells indicate that occupancy of the slow, high affinity receptors mediates this NGF function (*14, 39*). Moreover, slow receptors appear developmentally on chick sensory neurons only when the cells become responsive to NGF (*39*). On the other hand, stimulation of amino acid uptake in PC12 cells requires significantly higher concentrations of NGF than are necessary for promoting neurite outgrowth (*21, 31*). The retrograde flow of NGF also shows two receptor-mediated capacities and affinities reminiscent of the two classes of NGF receptors defined by binding studies (*9*). Such data suggest that the various functions of NGF may be subserved by different receptors and thus by different intracellular signals.

The molecular weights of the two classes of receptors have been determined by crosslinking ^{125}I-NGF to membranes or cells using a heterobifunctional, photoactivated crosslinking agent. On rabbit sympathetic neurons the NGF-NGF receptor complexes have molecular weights of 143 K and 112 K (*30*) while on PC12 cells the molecular weights are 158 K and 100 K (*19*). On the basis of their rates of dissociation, trypsin sensitivity and occupancy at different NGF concentrations Hosang and Shooter (*19*) were able to identify the 158 K complex as the slow and the 100 K complex as the fast NGF receptor complexes, respectively. The molecular weights of the slow and fast NGF receptors themselves are, therefore, approximately 140 K and 82 K, respectively. Studies on the biosynthesis of the human fast NGF receptor in melanoma A875 cells (13) are consistent with the idea that the fast receptor is a single peptide chain with a significant degree of glycosylation.

The structure of the rat fast NGF receptor has now been determined (33). The gene for the rat receptor was first transferred to and expressed in mouse L cells. In one of the transfectants, PCNA, the expression of the receptor was amplified through 10 cycles of cell sorting and growth in HAT medium until a stable line expressing approximately 2.5×10^6 receptors per cell was obtained. The cDNA for the fast NGF receptor was rescued from a PCNA cDNA library with an NGF receptor-enriched probe made by subtractive hybridization of PCNA cDNA with constitutive L cell RNA. The cDNA,

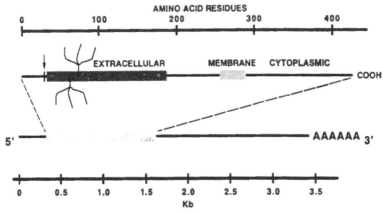

Fig. 2. The fast NGF receptor mRNA and its protein transcript. In the second line from the bottom the coding region in the mRNA is indicated by a box. It lies entirely in the 5' half of the mRNA. The open reading frame is expanded into the third line from the bottom to show diagrammatically the 425 amino acid precursor of the fast NGF receptor. The arrow near the left-hand end indicates where the 29 amino acid signal peptide is cleaved to produce the mature receptor protein. The box which begins immediately after the arrow shows the cysteine rich region in the extracellular domain. Within this region are two putative N-linked glycosylation sites indicated by trees. The single membrane spanning domain (22 amino acid residues) is indicated near the center of the sequence; it separates the 222 amino acid residue extracellular domain from the 152 amino acid residue cytoplasmic domain. The latter lacks an ATP binding site.

which is approximately 3.3 kb long was proven to be a fast NGF receptor cDNA by its ability to give rise to fast NGF receptor expression when it was cloned into the mammalian expression vector, pcDL1, and transfected into L cells. The fast NGF receptor pUC9 clone was called pNGFR1. The first start codon is 114 bases from the 5' end (Fig. 2). The open reading frame continues for 1,275 bases resulting in a predicted fast NGF receptor precursor of 425 amino acid residues in length. An N-terminal sequence for the purified receptor shows that 29 N-terminal amino acid residues are removed leaving a receptor peptide with 396 residues and a molecular weight of 42,478. Given the molecular weight of the mature receptor (82 K) it is clear that the fast NGF receptor is rich in carbohydrate which is presumably attached to one or both putative N-glycosylation sites or a number of putative O-glycosylation sites in the extracellular domain.

The receptor is unique at both nucleotide and amino acid sequence levels. It has one membrane spanning a domain of 22 residues dividing the receptor into a 222 residue long extracellular domain and a 151 residue long cytoplasmic domain. The extracellular domain comprises four repeating elements rich in cysteine residues similar to structures found in other growth factor (mitogen) receptors (Fig. 3). The intra-

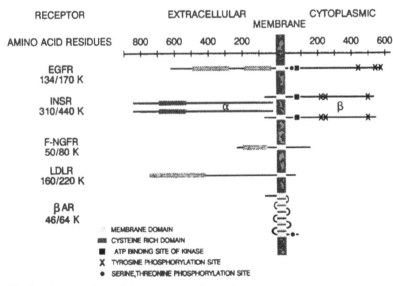

Fig. 3. A comparison of the structures of some known receptors.

cellular domain is too short to encode an endogenous tyrosine kinase activity and, in keeping with this, lacks the consensus sequence for an ATP binding site. A comparison with the structure of other receptors (Fig. 3) points up these similarities and differences. Since NGF is not a mitogen like, *e.g.* EGF, it is not surprising that its signal transduction mechanism does not involve an endogenous protein kinase. The LDL receptor is one of a class of receptors whose major function is the internalization of ligands. Yet it is not clear from looking at the two receptor structures what common feature would be responsible for this property.

Of considerable interest is the fact that the cDNA for the fast NGF receptor hybridizes to a single mRNA species even in cells (PC12 and sensory neurons) which express both types of receptor. This suggests that the two receptors share the protein coded for by the fast NGF receptor gene as a common NGF binding subunit. It follows that the slow NGF receptor must contain a second subunit, the effector protein, of approximately 60 K whose identity is, as yet, unknown. The alternative explanation that the two NGF receptors have no relationship other than their ability to bind NGF appears less likely but still must be borne in mind.

III. INTRACELLULAR PATHWAYS

The structure of the fast NGF receptors does not yet provide a definitive clue as to the signal transduction mechanism of NGF. Rather does it emphasize that the key to this mechanism lies in understanding the difference between the structures of the fast and slow receptors, *i.e.*, the nature of the effector protein. It is worth noting at this point that the internalization of NGF into PC12 cells is mediated by the slow but not the fast NGF receptor (*3, 38*), implying that the effector protein is a key element in this process. However, as pointed out earlier it is still not known whether internalization of NGF is part of the intracellular signalling mechanism.

The effects of NGF on proto-oncogene expression in PC12 cells may provide the clues to potential second and subsequent messengers. The transcription of the proto-oncogene *c-fos* is rapidly (within 10 min) increased while that of *c-myc* is also elevated but with a slower time course (*10*). Since these responses are not unique to NGF but are also induced by epidermal growth factor (EGF) and the tumor promoter TPA they represent general responses of PC12 cells to growth factor stimulation. *c-fos* codes for a 62 K nuclear phosphoprotein while *c-myc* also codes for a nuclear protein (*49*). Other changes in gene expression can be observed by following transcription-dependent changes in protein synthesis in the first few hours after exposure of PC12 cells to NGF. Interestingly, two nuclear matrix proteins of 56 K and 50 K, respectively are proteins whose synthesis is increased (*41*). Although no induction of these proteins was observed in dbcAMP-treated cells it is not yet known whether this response is a general one or specific to NGF.

If all the above changes are general responses to growth factors then there must be other specific genes which NGF induces or represses to bring about differentiation of the PC12 cells. Of interest here are the observations that the oncogenes *v-src* and *v-ras*, which result in non-regulated growth in cells that respond to EGF and PDGF, induce differentiation in PC12 cells in the absence of NGF (*1, 2, 32*). The *v-src*, in particular, codes for a membrane bound tyrosine kinase of 60 K raising the possibility that this protein might be the effector protein. Indeed, this protein has been shown to have a dual subcellular location in cells, being found in Golgi-like structures in the perinuclear region as well as in the plasma membrane of the cell (*34*). Such a distribution could result from its location on the plasma membrane and in internalized vesicles. It would also provide an explanation for the finding that NGF can induce the rapid phosphorylation of a number of proteins on tyrosine residues (*15*). Alternatively, the involvement of the *v-ras* product in the pathway of action of NGF may implicate one of the family of G proteins as the effector protein.

Other specific genes which are induced by NGF are those concerned with the synthetic enzymes for the neurotransmitters in PC12 cells (*42*) and the microtubule associated protein tau (and possibly MAP1) (*8*), a key regulator of microtubule and neurite stability. When PC12 cells are cultured with NGF for 7 days extensive neurite outgrowth occurs (Fig. 4a). The maintenance of the neurite network is dependent on the continued presence of NGF and this can be seen by the complete disintegration of neurites within 2 days of withdrawing NGF. As the PC12 cells grow neurites the amount of microtubules in

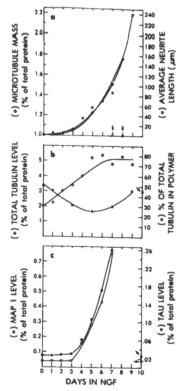

Fig. 4. Quantitative analysis of neurite length, microtubule mass, and microtubule protein levels during PC12 cell neurite extension. a: microtubule mass, determined by quantitative immunoblotting of 15 μg detergent-extracted cytoskeleton protein with antitubulin monoclonal antibodies, and average neurite length for 200 neurites measured each day, are plotted as a function of culture time in the presence of NGF. Arrows indicate data points collected after 2 days (O) or 3 days (●) of NGF withdrawal. b: total tubulin levels, determined by quantitative immunoblotting of 30 μg total PC12 cell protein with antitubulin monoclonal antibodies, are plotted as a function of culture time in NGF. The arrow marks a data point collected after 3 days of NGF withdrawal. The percentage of total tubulin in polymer form was determined from the ratio of microtubule mass in (a) to total tubulin level in (b). c: MAP1 and tau protein levels, determined by quantitative immunoblotting of 30 μg total PC12 cell protein with an MAP1 monoclonal antibody and an affinity-purified anti-tau serum, are plotted as a function of culture time in the presence of NGF. (Tau quantitation is for 61-, and 68-, and 125-K polypeptides combined.) Arrows indicate data points collected after 3 days of NGF withdrawal. Reproduced with permission from *J. Cell Biol.* (1985), **101**, 1799–1807.

the cell also increases and the time courses of the two processes essentially coincide (Fig. 4a). The microtubules which form in PC12 cells in the presence of NGF do so from a large pool of unpolymerized tubulin. That the maintenance of the neurite is dependent on intact microtubules is shown by the effects of microtubule depolymerizing drugs such as vinblastine or colchicine. Within 2 hr both neurites and microtubules completely disintegrate in the presence of these agents. Conversely, drugs such as taxol which stabilize microtubules preserve neurites.

Several mechanisms could act to increase microtubule content as the PC12 cells differentiate. Using a Western blot procedure and antibodies for tubulin or the microtubule associated proteins, MAP1 and tau, the concentrations of these components of microtubules have been measured during the differentiation of the PC12 cells. NGF causes an immediate increase in tubulin synthesis which persists for about 6 days before leveling off. However, this induction of tubulin precedes the formation of new microtubules and, indeed as Fig. 4b shows, tubulin levels increase much faster in the first 3 days than do microtubule levels. Clearly, tubulin accumulation does not drive microtubule assembly. In contrast, not only do the levels of MAP1 and tau increase significantly (20-fold and 10-fold, respectively) but the time course of these increases parallels those of neurite outgrowth and microtubule formation (Fig. 4c). Moreover, when NGF is withdrawn from the differentiated PC12 cells, MAP1 and tau levels fall rapidly to those found in undifferentiated cells as both neurites and microtubules disappear. Thus MAP1 and tau are limiting factors for microtubule assembly during neurite growth.

The quantitation of the molar ratios of MAP1 and tau in the assembled microtubules in PC12 cells shows that they are close to the saturating ratios found *in vitro* for very stable microtubules. Varying the conditions under which neurite outgrowth occurs adds credence to the idea that MAP1 and tau levels are the critical determinants of neurite outgrowth. When dbcAMP is added along with NGF neurite outgrowth is accelerated by about 2 days. The time courses of MAP1 and tau induction now follow this accelerated time course. Neurites grown in the presence of dbcAMP alone are short and unstable. Under these conditions tau is induced more than MAP1. All these results point to the key role of the microtubule associated proteins in microtubule and thus neurite assembly and stabilization. The flow of NGF from target to cell body ensures that appropriate levels of these proteins are maintained. It remains to be determined how NGF regulates the synthesis of these components and how this fits in the as yet unknown signal transducing mechanism of NGF.

SUMMARY

Nerve growth factor is responsible for the regulation of the selective survival of sympathetic and some sensory neurons during development. The key to this regulation is the flow of very small amounts of NGF from the targets of these nerve cells to their cell bodies, a flow which not only mediates survival but maintains the differentiated state of the neurons and promotes axonal regrowth after damage. During the flow intracellular signals are generated which activate the molecular pathways behind these functions of NGF. The structure of the fast NGF receptor, determined from the nucleotide sequence of its cDNA provides few clues about the nature of potential second messengers, except to emphasize that the latter are probably generated by the effector protein with which this receptor interacts after binding NGF. It is clear, however, that NGF promotes neurite (axonal) growth and stability by its role in the assembly and stabilization of microtubules and that it does this by the induction of the microtubule associated proteins, at least in part at the level of transcription.

Acknowledgment

Original work described in this paper was supported by grants from NINCDS (NS 04270), the American Cancer Society (BC 325) and the Isabella M. Niemela Fund.

REFERENCES

1 Alema, S., Caralbove, P., Agostini, E., and Tats, F. (1985). *Nature* **316**, 557–559.
2 Bar-Sagi, D. and Feramisco, J. (1985). *Cell* **42**, 841–848.
3 Bernd, P. and Greene, L.A. (1984). *J. Biol. Chem.* **259**, 15509–15516.
4 Block, T. and Bothwell, M. (1983). *J. Neurochem.* **40**, 1654–1663.
5 Buxser, S.E., Kelleher, D.J., Watson, L., Puma, P., and Johnson, G.L. (1983). *J. Biol. Chem.* **258**, 3741–3749.

6 Cohen, S. (1960). *Proc. Natl. Acad. Sci. U.S.A.* **46**, 302–311.

7 Davies, A., Bandtlow, C., Heumann, R., Korsching, S., Rohrer, H., and Thoenen, H. H. (1986). *Soc. Neurosci. Abstr.* **12**, 1093.

8 Drubin, D.G., Feinstein, S.C., Shooter, E.M., and Kirschner, M.W. (1985). *J. Cell Biol.* **101**, 1799–1807.

9 Dumas, M., Schwab, M.E., and Thoenen, H. (1979). *J. Neurobiol.* **10**, 179–197.

10 Greenberg, M.E., Greene, L.A., and Ziff, E.B. (1985). *J. Biol. Chem.* **260**, 14101–14110.

11 Greene, L.A. and Tischler, A.S. (1976). *Proc. Natl. Acad. Sci. U.S.A.* **73**, 2424–2428.

12 Grob, P.M. and Bothwell, M.A. (1983). *J. Biol. Chem.* **258**, 4136–4143.

13 Grob, P.M., Ross, A.H., Koprowski, H., and Bothwell, M.A. (1985) *J. Biol. Chem.* **260**, 8044–8049.

14 Gunning, P.W., Landreth, G.E., Bothwell, M.A., and Shooter, E.M. (1981). *J. Cell Biol.* **89**, 240–245.

15 Halegoua, S. and Patrick, J. (1980). *Cell* **22**, 571–581.

16 Hamburger, V. and Levi-Montalcini, R. (1949). *J. Exp. Zool.* **111**, 457–501.

17 Hendry, I.A. (1977). *Brain Res.* **134**, 213–223.

18 Heumann, R., Korsching, S., Scott, J., and Thoenen, H. (1984). *EMBO J.* **3**, 3183–3189.

19 Hosang, M. and Shooter, E.M. (1985). *J. Biol. Chem.* **260**, 655–662.

20 Johnson, E.M., Gorin, P.D., Brandeis, L.D., and Pearson, J. (1980). *Science* **210**, 916–918.

21 Kedes, D.H., Gunning, P.W., and Shooter, E.M. (1982). *J. Neurosci. Res.* **8**, 357–365.

22 Korsching, S.I., Heumann, R., Davies, A., and Thoenen, H. (1986). *Soc. Neurosci. Abstr.* **12**, 1096.

23 Korsching, S.I. and Thoenen, H. (1983). *Neurosci. Lett.* **39**, 1–4.

24 Korsching, S. and Thoenen, H. (1985). *J. Neurosci.* **5**, 1058–1061.

25 Landreth, G.E. and Shooter, E.M. (1980). *Proc. Natl. Acad. Sci. U.S.A.* **77**, 4751–4755.

26 Levi-Montalcini, R. and Angeletti, P.U. (1966). *Pharmacol. Rev.* **48**, 534–569.

27 Levi-Montalcini, R. and Hamburger. V. (1951). *J. Exp. Zool.* **116**, 321–362.

28 Levi-Montalcini, R. and Hamburger, V. (1953). *J. Exp. Zool.* **123**, 233–278.

29 Levi-Montalcini, R., Meyer, H., and Hamburger, V. (1954). *Cancer Res.* **14**, 49–57.

30 Massague, J., Guillett, B.J., Czech, M.P., Morgan, C.J., and Bradshaw, R.A. (1981). *J. Biol. Chem.* **256**, 9419–9424.

31 McGuire, J.C. and Greene, L.A. (1979). *J. Biol. Chem.* **254**, 3362–3367.

32 Noda, M., Ko. M., Ogura, A., Lim, D.-G., Amano. T., Takano, T., and Ikawa, Y. (1985). *Nature* **318**, 73–75.

33 Radeke, M.J., Misko, T.P., Hsu, C., Herzenberg, L.A., and Shooter, E.M. (1987). *Nature* **325**, 593–597.

34 Resh, M.D. and Erickson, R.L. (1985). *J. Cell Biol.* **100**, 409–417.

35 Schechter, A.L. and Bothwell, M.A. (1981). *Cell* **24**, 867–874.

36 Scott, J., Selby, M., Urdea, M., Quiroga, M., Bell, G.I., and Rutter, W.J. (1983). *Nature* **302**, 538–540.

37 Shelton, D.L. and Reichardt, L.F. (1984). *Proc. Natl. Acad. Sci. U.S.A.* **81**, 7951–7955.

38 Shooter, E.M., Yankner, B.A., Landreth, G.E., and Sutter, A. (1981). *Recent Prog. Horm. Res.* **37**, 417–446.

39 Sutter, A., Riopelle, R.J., Harris-Warrick, R.M., and Shooter, E.M. (1979). In *Transmembrane Signaling,* eds. Bitensky, M., Collier, R.J., Steiner, D.F., and Fox, C.F., pp. 659–677. New York: Alan R. Liss Inc.

40 Sutter, A., Riopelle, R.J., Harris-Warrick, R.M., and Shooter, E.M. (1979). *J. Biol. Chem.*

254, 5972–5982.

41 Tiercy, J.-M. and Shooter, E.M. (1986). *J. Cell Biol.* **103**, 2367–2378.

42 Thoenen, H. and Barde, Y.A. (1980). *Physiol. Rev.* **60**, 1284–1335.

43 Ullrich, A., Gray, A., Berman, C., and Dull, T.J. (1983). *Nature* **303**, 821–825.

44 Vale, R.D., Ignatius, M.J., and Shooter, E.M. (1985). *J. Neurosci.* **5**, 2762–2770.

45 Vale, R.D. and Shooter, E.M. (1982). *J. Cell Biol.* **94**, 710–717.

46 Vale, R.D. and Shooter, E.M. (1983). *Biochemistry* **22**, 5022–5028.

47 Vale, R.D. and Shooter, E.M. (1985). *Methods Enzymol.* **109**, 21–39.

48 Varon, S., Nomura, J., and Shooter, E.M. (1967). *Proc. Natl. Acad. Sci. U.S.A.* **57**, 1782–1789.

49 Verma, I.M. (1984). *Nature* **308**, 317.

4

MOLECULAR GENETIC ANALYSIS OF THE MYELIN DEFICIENT MUTANT ANIMALS

KATSUHIKO MIKOSHIBA,*1 HIDEYUKI OKANO,*1
MASAYUKI MIURA,*1 KAZUHIRO IKENAKA,*1
YASUZO TSUKADA,*2 AND YOSHIRO INOUE*3

Division of Regulation of Macromolecular Function, Institute for Protein Research, Osaka University, Suita, Osaka 565,* Department of Physiology, School of Medicine, Keio University, Tokyo 160,*2 and Department of Anatomy, Faculty of Medicine, Hokkaido University, Sapporo 060, Japan

Myelin is a structure that facilitates conduction of nerve impulses along the axon. It is formed by oligodendrocytes in the central nervous system (CNS), and by Schwann cells in the peripheral nervous system (PNS). Oligodendrocytes originate from the neural tube and Schwann cells originate from neural crest cells. The molecular architecture of myelin and the way of myelination in CNS differ from those in PNS. In the CNS, several processes of an oligodendrocyte form myelins, while in the PNS the Schwann cell body itself surrounds the axon to form a compact myelin.

Recently many neuropathological mutants with abnormal myelination have been reported, such as *shiverer, myelin deficient (mld), jimpy, quaking, myelin synthesis deficiency, trembler, and twitcher.* Analyses of these mutants compared with respective wild type control animals should give us good information on the role of myelin components in myelination.

In this article, we shall present recent studies on myelination analyzing the *shiverer* mutant mouse and its allelic mutant, *myelin deficient (mld)*.

TABLE I

Myelin Deficient Mutant Mice (*shiverer* and *mld*)

1) Autosomal recessive
   Chromosomal localization (*18*)
   Allelic mutation ($shi^{=ld}$)
2) Symptoms
   Intentional tremor
   Tonic convulsion
   Ataxic movement
3) Abnormal and poor myelination

I. MORPHOLOGICAL AND BIOCHEMICAL STUDIES ON *SHIVERER* MUTANT

When there is an abnormality in myelination, the animal shows abnormal behavior such as tremor in both trunk and extremities, tonic convulsions, and/or ataxic movement. Since similar abnormal behaviors are found in patients afflicted with the human disease, leukodystrophy, these mutants are used as its animal model. *Shiverer* mouse was discovered as a myelin deficient animal (*1*). When we observed the brain of the *shiverer* mutant animals by electron microscopy, there were clear differences in the myelin: the myelin lamella was much thinner than that of the control (*7, 8, 37*), and there was a greater number of immature type oligodendrocytes (*22*). The major dense line, which is formed by fusion of the inner surface membrane of oligodendrocytes, was absent (Fig. 3) (*37*), and the myelin lamella formation pattern was also abnormal (Fig. 3) (7). Some of the axons were non-myelinated or very loosely myelinated in the immature form. The most common observation was that myelin was formed by piling up the cytoplasmic sheets of oligodendrocytes (Fig. 4). Since a cytoplasmic sheet cannot surround the axon, the other processes from another oligodendrocyte or the same oligodendrocyte take place to wrap the piled processes again. This piling up of the cytoplasmic sheets of oligodendrocytes results in the abnormal myelin lamella where major dense lines are absent in the *shiverer* mutant mice (*7–9*). Not only were there abnormalities in the myelin structure of the *shiverer* CNS, but there were also abnormalities in other cells: hypertrophy of the astrocytic processes (*32*), and an abnormal fiber arborization of neurons.

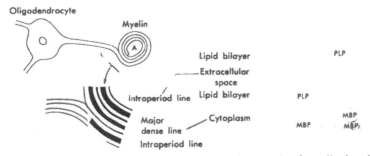

Fig. 1. Brief scheme of electron microscopic photograph of myelin lamella and its molecular architecture. PLP, proteolipid protein; MBP, myelin basic protein.

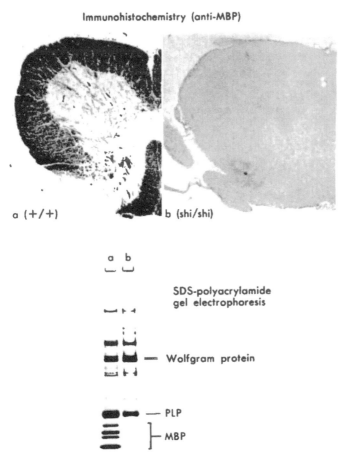

Fig. 2. Immunohistochemical reaction against myelin basic protein on the sections from wild type control ($+/+$) and *shiverer* mutant mice (*shi/shi*) with gel electrophoretic pattern of the myelin fractions from respective animals. From Mikoshiba *et al.* (*22*).

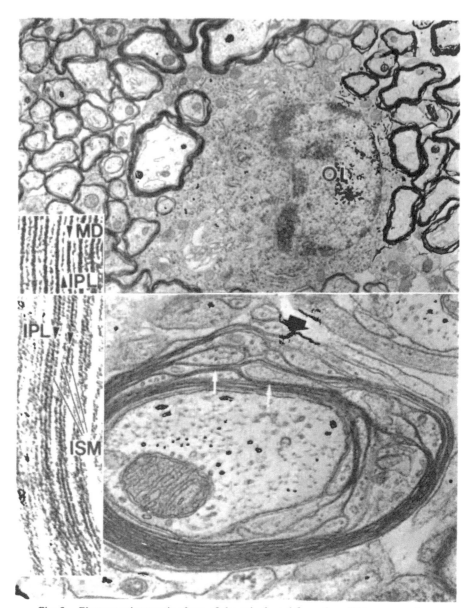

Fig. 3. Electron microscopic photo of the spinal cord from the wild type control mouse and the *shiverer* mutant mouse. Upper: cross section from the wild type control mouse. Lower: cross section from the *shiverer* mutant mouse. IPL, intraperiod line; ISM, inner surface membrane of oligodendrocyte; OL, oligodendrocyte. From Inoue *et al.* (7).

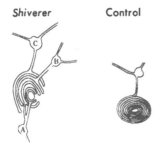

Fig. 4. Schematic drawing of the typical *shiverer* myelin lamella.

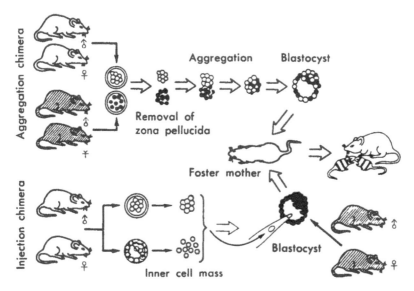

Fig. 5. Scheme for production of aggregation chimera mouse. From Mikoshiba *et al.* (25).

Analysis of the protein components of myelin in *shiverer* revealed several abnormalities. Severe loss of myelin basic protein (MBP) has been shown immunohistochemically using the antiserum against MBP (3, 22) or by SDS-gel electrophoresis (20, 21). Decrease in proteolipid protein was also found by gel electrophoresis (20, 21). We did not know at the time of our experiments what caused the myelin abnormality in the *shiverer*, nor did we know the primary site of the mutation at the cellular level. Since the nervous system shows prominent compensatory effect among the cells, it was necessary to determine what cells were primarily affected in the *shiverer* nervous system.

To study this, we produced chimeras composed of *shiverer*-derived cells and normal animal-derived cells. One of the ways to produce chimeras is by aggregation (Fig. 5). Analysis of the chimera brain makes it possible to see the interactions between the normal and mutant cells. The interaction might be through cell surface to cell surface communication, or through communication mediated by the humoral factors, either toxic or trophic.

II. ANALYSIS OF CELLS PRIMARILY AFFECTED BY THE MUTATION BY PRODUCTION OF CHIMERAS

In order to judge the mosaicism, we chose mice with a different isozyme pattern and coat color than *shiverer* mutant mice. C57Bl (black coat

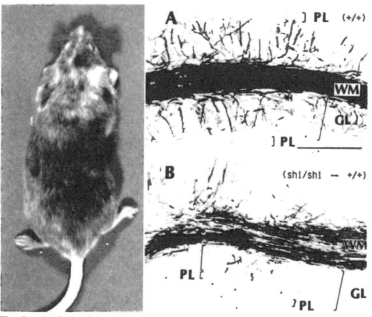

Fig. 6. Analysis of the brain from a chimera mouse produced by aggregation of 8-cell-stage embryo obtained from *shiverer* and wild type control mice. Left: the chimera mouse in which cerebellum was analyzed by immunohistochemistry using antiserum against myelin basic protein. Right: the sagittal section of the cerebellum from the control (A) mouse and the chimera mouse (B) shown in the left figure. PL, Purkinje cell layer; WM, white matter; GL, granular layer. From Mikoshiba *et al.* (*24, 26*).

color) mice thus chosen have GPI-1 isozyme pattern b, while *shiverer* has GPI-1 on a genetic background BALB/C (white coat color). As a tentative marker to identify the origin of myelin, we used the absence of MBP reaction and/or absence of major dense line as a marker to identify *shiverer* derived myelin. Figure 6 shows one of the chimeras obtained.

We have examined the GPI isozyme pattern of part of the brain, and found that the chimera brain is composed of both normal and *shiverer* derived cells (*24*). We could clearly observe the MBP positive and negative sites in the white matter of the brain (Fig. 6) (*26*). If some toxic factor existing in *shiverer* causes the formation of abnormal myelin in CNS, the chimera brain should be entirely *shiverer* type. And if the trophic factor that possibly exists in normal mice is missing in the *shiverer*, all the white matter of the chimera should be normal type. Patchy figures observed have eliminated the circulating humoral factor that causes the abnormal myelination. Thus the *shiverer* mutation must be derived from the intrinsic nature of the cells in either myelin forming cells or neuronal cells. If the mutation occurs on the neuronal side, normal axon would always have normal myelin and abnormal (*shiverer*-derived) axon would always have abnormal myelin. And if

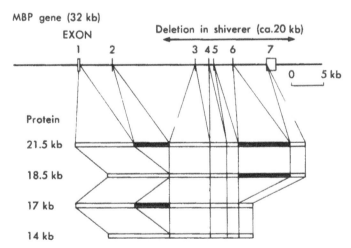

Fig. 7. Configuration of myelin basic protein gene and the myelin basic protein. This figure was constructed from data from Kimura *et al.* (*11*) Molineaux *et al.* (*35*), Takahashi *et al.* (*42*), de Ferra *et al.* (*4*), and Roach *et al.* (*39*).

Fig. 8. Immunohistochemical reaction on the sections from the sciatic nerve fiber from the chimera mouse. Upper: cross section. Lower: sagittal section. MBP-positive fiber is derived from the wild type mouse and MBP-negative fiber is derived from the *shiverer*. From Mikoshiba *et al.* (*29*).

normal myelin is found adjacent to *shiverer* type myelin, there should be no abnormality on neuronal side. As shown in Fig. 9, we observed normal myelin and abnormal myelin on one axon, and therefore concluded that mutation exists in myelin forming cells (*24*). This was also confirmed by transplantation of isolated oligodendrocytes or tissue blocks containing oligodendrocytes in *shiverer* brain (*13, 15*).

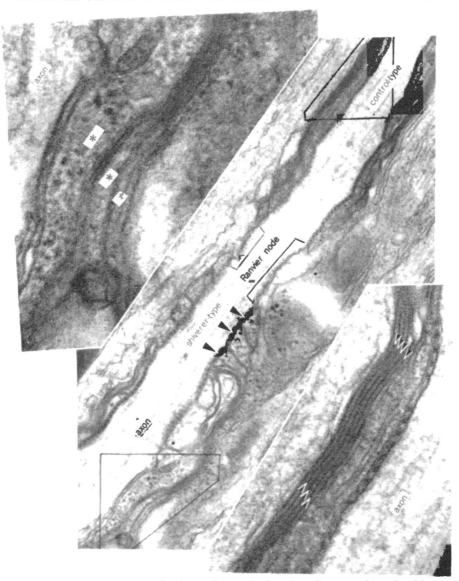

Fig. 9. Electron microscopic photo of the sagittal section of an axon from the chimera mouse brain. In the typical myelin lamella, major dense lines (indicated by small arrowheads) are clearly observed (lower inset) and paranodal cytoplasmic pockets lose their cytoplasm immediately to form the major dense line (central figure), while in the myelin of the lower left in the central figure and upper inset, which is believed to be of *shiverer* origin, the paranodal cytoplasmic pockets are enlarged and irregular in form and isolated pockets (indicated by large arrowheads) are often observed. In the *shiverer*-type myelin, cytoplasm of the oligodendrocyte (indicated by *) is found where the major dense line would form in normal myelin. From Mikoshiba *et al.* (*24*).

These results indicated that the abnormal arborization process of neurons and also the hypertrophy of astrocytes are secondary phenomena resulting from abnormal myelination. This was the moment at which we were able to correlate directly the absence of MBP with the abnormality in oligodendrocyte (28). MBP was found to be absent not only in the CNS but also in the PNS in the *shiverer* mice by gel electrophoresis (12, 23, 27) or immunohistochemically (23, 27). Contrary to CNS, PNS myelin in the *shiverer* showed compact lamella very similar to that of the control (23, 27, 29). Since there is so little MBP in PNS myelin that MBP does not contribute much for formation of compact myelin lamella. As with CNS, MBP-positive myelin was clearly observed adjacent to MBP-negative myelin on a nerve fiber (Fig. 8) (29). We obtained the same conclusion in the PNS, namely, the myelin forming cell is the primary mutation site in the *shiverer*.

III. MOLECULAR GENETIC STUDY ON *SHIVERER* MUTANT

The chimeric study on mutants has opened a new field of molecular genetic analysis of myelination. Analysis of MBP gene in *shiverer* to test whether or not the MBP gene exists has begun. Since MBP is well known as a protein that causes experimental allergic encephalitis (EAE) in an animal when injected, most of the amino acid sequences have been analyzed in rat, bovine, and human (2). However, the amino acid sequence of mouse MBP has not been reported. Common amino acid sequences were chosen to synthesize oligonucleotides as probes to clone the mouse MBP cDNA. Hirose synthesized the probes with which Kimura *et al.* (10) successfully cloned the cDNA from the mouse cDNA library. By Southern blot hybridization with that clone, Kimura and his colleagues found that most of the MBP genes are deleted in *shiverer* (10). The same result was observed by Roach *et al.* using rat MBP cDNA clone as a probe (39). Normal MBP gene was found to be composed of 7 exons. The four forms of myelin basic proteins were found to be synthesized by alternative splicing of a precursor transcribed from a single gene (Fig. 7), and MBP genes were mapped on chromosome No. 18 (39–41). The length of the MBP genomic gene was 30 kb, in which the third to seventh exons including introns were deleted in the *shiverer* making it about 20 kb long. *Shiverer* MBP gene still contains the first and second exons, with an intact promoter region of the MBP. Miura *et al.* tested whether the first exon is transcribed or further translated in *shiverer* (*31*). By oligonucleotide directed site specific mutagenesis, *Hind*III site was introduced into the first exon of MBP cDNA, and anti-sense RNA was prepared *in vitro* by SP6 RNA polymerase after subcloning into pSP64. With this first exon specific probe it was found that a considerable amount of MBP-RNA was transcribed in the *shiverer* mice. In order to study the expression at the protein level, peptides corresponding to the N-terminal 16 amino acids of MBP were synthesized by peptide synthesizer. The peptides synthesized were injected into rabbit to raise the antibody. However, no reaction was observed with this antibody in the *shiverer* white matter section, while strong positive reaction was observed in the control.

The fact that the first exon of MBP gene was transcribed indicated that promoter activity of the gene is normal, but there might be an abnormal translation step, probably due to the instability of mRNA that lacks poly(A)$^+$ tail as a result of the absence of the poly(A)$^+$ additional signal in *shiverer*. These experiments revealed that *shiverer* abnormality is simply caused by the deletion of the MBP gene.

IV. *MLD*, ALLELIC MUTANT TO *SHIVERER*

While we were investigating *shiverer* mutation, we came across an interesting mutant, '*myelin deficient*' (*mld*), which is allelic to *shiverer* (*14*). By intercrossing homozygous *mld* and *shiverer*, we obtained 100% symptomatic offspring (Table II). The analysis of the *mld/shi* mice showed that the MBP was expressed, but at less than the level of *mld* mice. Matthieu *et al.* have extensively studied *mld* mutant mice and showed that the major dense line was almost absent, but MBP was slightly produced (*16–19, 38*). Their observation stimulated us to analyze the *mld* brain as we did in *shiverer* (Figs. 10–12). We found that MBP was expressed in a mosaic manner with regard to immunohistochemical reaction and electron microscopic observation (Figs. 11 and 12). Some among the various myelins showed a normal lamella with major dense line, indicating that MBP gene was fully expressed. Other myelins

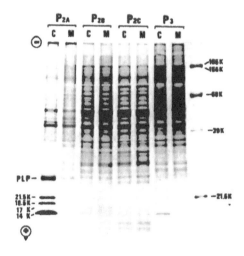

Fig. 10. Protein profiles of the subcellular fractions from the brain of wild type control and *mld* mice. C, fractions from wild type control; M, fraction from *mld* mutant mice; P2A, myelin fraction; P2B, synaptosomal fraction; P2C, mitochondrial fraction; P3, microsomal fraction; PLP, proteolipid protein. From Mikoshiba *et al.* (*30*).

TABLE II

Allelism Test between *shiverer* and *mld* Mutants

Parents	Symptomatic offspring/total offspring
mld/mld × *shi/shi*	24/24

mld/mld: MDB/dt. *shi/shi*: BALB/c

showed no MBP reaction and no major dense line. Yet others showed slight MBP reaction and a few major dense lines in a single myelin lamella (*30*).

V. MOLECULAR GENETIC STUDY ON *MLD* MUTANT

By RNA dot blot analysis of *mld* mouse brain, it was shown that mRNA for MBP is synthesized at about 3% of the control value. Northern blot analysis revealed that the size of RNA synthesized was the same as that of the control. Southern blot analysis of the *mld* genomic DNA cleaved by various restriction enzymes revealed that the pattern was similar to that of the control using mouse MBP-cDNA as a probe. However, when we used different restriction enzymes and different probes, we found a duplication of bands, indicating that there is a

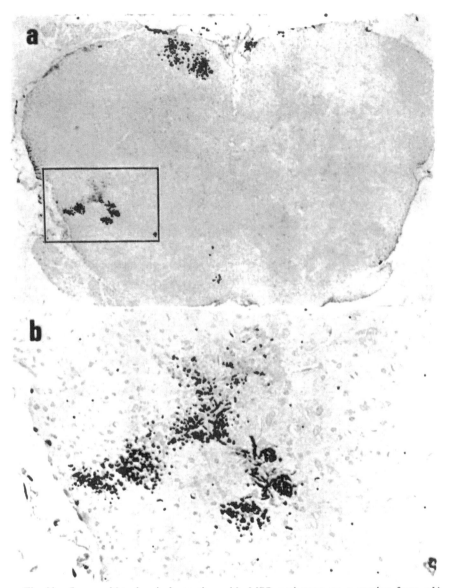

Fig. 11. Immunohistochemical reaction with MBP antiserum on a section from *mld* mutant spinal cord. MBP-positive reactions are seen as patches. b is a higher magnification of the left part of the spinal cord in a. Intensity of the reaction differs from myelin to myelin. From Mikoshiba *et al.* (*30*).

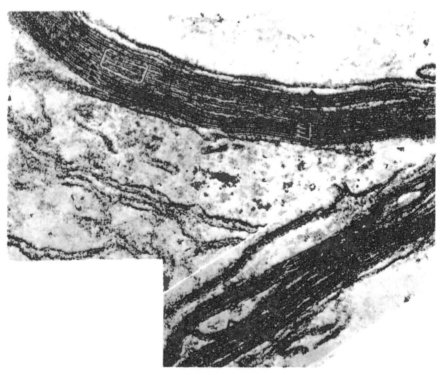

Fig. 12. Electron microscopic figure of the *mld* mutant mouse. High density of the major dense line is indicated by small arrowheads. In the upper figure, the lamella appears compact but the density of the major dense line differs from line to line, and the absence of a major dense line in one area is indicated by the large open arrowhead. Myelin lamella where major dense line is absent is also indicated by parentheses. In the lower figure, cytoplasmic loop is indicated by large open arrowhead. From Mikoshiba *et al.* (*30*).

duplication in MBP gene in *mld* (*33, 34*) (Fig. 14). The duplication of the gene might cause the inefficient transcription of MBP gene. We succeeded in cloning the two types of promoter regions of the MBP genes in *mld* mutant mice (Fig. 15), and are actively investigating the mechanism of transcriptional interference of the MBP gene expression in detail.

CONCLUSION

In conclusion, neuropathological mutants have opened a new field of molecular genetic study in neurobiology. Studies on *shiverer* and *mld*

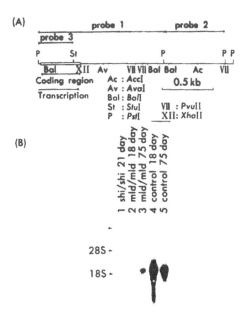

Fig. 13. A: restriction map of the MBP cDNA clone from mouse. From Kimura *et al.* (*10*). The open box is the coding region, the length of which is 381 bp. The direction of transcription is from left to right. Probes used in Fig. 14 are indicated as probes 1, 2, 3. B: Northern blot analysis of RNAs from the brains of *shiverer*, *mld*, and wild type control mice. Mouse MBP cDNA was used as a probe. Total RNA (20 μg) from brains of *shiverer* (lane 1, day 21), *mld* (lane 2, day 18; lane 3, day 75) and control (lane 4, day 18; lane 5, day 75) was electrophoresed on 1% agarose gel. From Okano *et al.* (*34*).

mutant mice in particular have given us a great amount of information and will give us more on the mechanism of gene expression of MBP and also on the role of MBP in myelination.

SUMMARY

We here report molecular genetic studies on myelin deficiency in *shiverer* and *mld* mice.

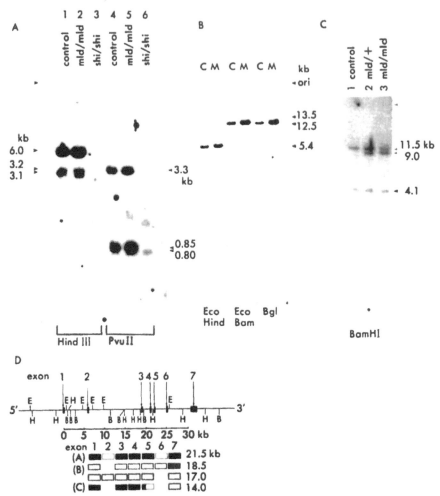

Fig. 14.   Southern blot analysis of DNAs from mice using mouse MBP cDNA as a probe. Ten micrograms of liver chromosomal DNA from the mice indicated was digested with the restriction enzymes indicated. For the hybridization, probe-1 and probe-2 were used in A, probe-2 was used in B, and probe-3 was used in C. Probes used in the experiment are indicated in Fig. 13. Region of the probes used in A, B, C are indicated as closed boxes in D. From Okano *et al.* (*34*).

In *shiverer* mice, the MBP content was absent in both CNS and PNS. The MBPmRNA was not detectable, although a considerable amount of RNA was transcribed from the MBP gene. By Southern blot studies the

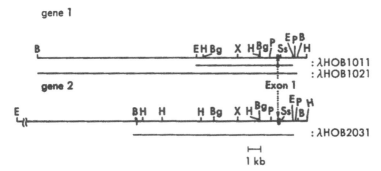

Fig. 15. Structure of 5'-region of the two types of myelin basic protein gene in *mld* mutant mice. B, *Bam*HI; E, *Eco*RI, H, *Hind*III; Ss, *Sst*I.

MBP genomic gene was found to be deleted, and this deletion resulted in the absence of MBP in the *shiverer*.

However, *mld* mutant, allelic to *shiverer*, showed a slight expression of MBP detected by Western blotting and immunohistochemistry. The level of mRNA synthesized in *mld* was decreased to about 3% of the control although the size of the RNA was the same as control. Southern blot study showed duplication of the MBP gene in *mld* mice, which resulted in the inefficient transcription of MBP.

REFERENCES

1 Biddle, R., March, E., and Miller, J.R. (1973). *Mouse News Lett.* **48**, 24.

2 Dunkley, P.R. and Carnegie, P.R. (1974). *Biochem. J.* **141**, 243–255.

3 Dupouey, P., Jacque, C., Bourre, J.M., Cesselin, F., and Privat, A. (1979). *Neurosci. Lett.* **12**, 113–118.

4 de Ferra, F., Engh, H., Hudson, L., Kamholz, J., Puckett, C., Molineaux, S., and Lazzarini, R.A. (1985). *Cell* **43**, 721–727.

5 Hogan, E.L. and Greenfield, S. (1984). In *Myelin*, ed. Morell, P., 2nd ed., pp. 489–534. New York and London: Plenum Press.

6 Inoue, Y., Mikoshiba, K., Yokoyama, M., Inoue, K., Terashima, T., Nomura, T., and Tsukada, Y. (1986). *Dev. Brain Res.* **26**, 239–247.

7 Inoue, Y., Nakamura, R., Mikoshiba, K., and Tsukada, Y. (1981). *Brain Res.* **219**, 86–94.

8 Inoue, Y., Inoue, K., Terashima, T., Mikoshiba, K., and Tsukada, Y. (1983). *Anat. Embryol.* **168**, 159–171.

9 Inoue, Y., Nakamura, R., Mikoshiba, K., and Tsukada, Y. (1982). *Okajima Folia Anat. Japon.* **58**, 613–625.

10 Kimura, M., Inoko, H., Katsuki, M., Ando, A., Sato, T., Hirose, T., Takashima, H., Inayama, S., Okano, H., Takamatsu, K., Mikoshiba, K., Tsukada, Y., and Watanabe, I. (1985). *J. Neurochem.* **44**, 692–696.

11 Kimura, M., Katsuki, M., Inoko, H., Ando, A., Sato, T., Hirose, T., Inayama, S., Taka-

shima, H., Takamatsu, K., Mikoshiba, K., Tsukada, Y., Yokoyama, M., and Watanabe, I. (1986). In *Molecular Genetics in Developmental Neurobiology*, ed. Tsukada, Y., pp. 125–136. Tokyo and Utrecht: Japan Sci. Soc. Press and VNU Science Press BV.

12 Kirschner, S.A. and Ganser, A.L. (1980). *Nature* 283, 207–210.

13 Kohsaka, S., Yoshida, K., Inoue, Y., Shinozaki, T., Takayama, H., Inoue, H., Mikoshiba, K., Takamatsu, K., Otani, M., Toya, S., and Tsukada, Y. (1986). *Brain Res.* 372, 137–142.

14 Lachapelle, F., Baecque, C.De, Jacque, C., Bourre, J.M., Delassalle, A., Doolittle, D., Hauw, J.J., and Baumann, N. (1980). In *Neurological Mutations Affecting Myelination*, ed. Baumann, N., pp. 27–32. Amsterdam: Elsevier/North-Holland Biomed. Press.

15 Lachapelle, F., Gumpel, M., Baulac, M., Jacque, C., Duc, P., and Baumann, N. (1984). *Dev. Neurosci.*, 6, 325–334.

16 Matthieu, J.-M., Ginalski, H., Friede, R.L., Cohen, S.R., and Doolittle, D.P. (1980). *Brain Res.* 191, 278–283.

17 Matthieu, J.-M., Ginalski, H., Friede, R.L., and Cohen, S.R. (1980). *Neuroscience* 5, 2315–2320.

18 Matthieu, J.-M., Herschkowitz, N., Kohler, R., and Heitz, P.U. (1983). *Brain Res.* 268, 267–274.

19 Matthieu, J.-M., Omlin, F.X., Ginalski, H., and Cooper, B.JH. (1984). *Dev. Brain Res.* 13, 149–158.

20 Mikoshiba, K., Aoki, E., and Tsukada, Y. (1980). *Brain Res.* 192, 195–204.

21 Mikoshiba, K., Nagaike, K., and Tsukada, Y. (1980). *J. Neurochem.* 35, 465–470.

22 Mikoshiba, K., Takamatsu, K., Kohsaka, S., Tsukada, Y., and Inoue, Y. (1982). In *Genetic Approaches to Developmental Neurobiology*, ed. Tsukada Y. pp. 195–221. Tokyo and Berlin: Univ. Tokyo Press and Springer.

23 Mikoshiba, K., Kohsaka, S., Takamatsu, K., and Tsukada, Y. (1981). *Brain Res.* 204, 455–460.

24 Mikoshiba, K., Yokoyama, M., Inoue, Y., Takamatsu, K., Tsukada, Y., and Nomura, T. (1982). *Nature* 229, 357–359.

25 Mikoshiba, K. (1982). In *Cell Sociology in the Chimera*, ed. Kato, Y. and Mikoshiba, K., pp. 73–85. Tokyo: Kodansha Scientific.

26 Mikoshiba, K., Yokoyama, Y., Takamatsu, K., Tsukada, Y., and Nomura, T. (1982). *Dev. Neurosci.* 5, 520–524.

27 Mikoshiba, K., Takamatsu, K., and Tsukada, Y. (1983). *Dev. Brain Res.* 7, 71–79.

28 Mikoshiba, K., Yokoyama, M., Inoue, Y., Takamatsu, K., Nomura, T., and Tsukada, Y. (1984). *Experimental Allergic Encephalomyelitis: A Useful Model for Multiple Sclerosis*, pp. 475–480. New York: Alan R. Liss.

29 Mikoshiba, K., Yokoyama, M., Takamatsu, Y., and Nomura, T. (1984). *Dev. Biol.* 105, 221–226.

30 Mikoshiba, K., Okano, H., Fujishiro, M., Inoue, Y., Takamatsu, K., Lachapelle, F., Baumann, N., and Tsukada, Y. (1987). *Dev. Brain Res.*, 35, 111–121.

31 Miura, M., Ikenaka, K., Okano, H., Shioda, C., Tsukada, Y., and Mikoshiba, K. (1986). *Bull. Jpn. Neurochem. Soc.* 25, 574–576.

32 Nagaike, K., Mikoshiba, K., and Tsukada, Y. (1982). *J. Neurochem.* 39, 1235–1241.

33 Okano, H., Miura, M., Mikoshiba, K., Inoue, Y., Fujishiro, M., and Tsukada, Y. (1986). In *Molecular Genetics in Developmental Neurobiology*, ed. Tsukada, Y., pp. 187–199. Tokyo and Utrecht: Japan Sci. Soc. Press and VNU Science Press BV.

34 Okano, H., Miura, M., Moriguchi, A., Ikenaka, K., Tsukada, Y., and Mikoshiba, K.

(1987). *J. Neurochem.* **48**, 470–477.

35 Molineaux, S.M., Engh, H., de Ferra, F., Hudson, L., and Lazzarini, R.A. (1986). *Proc. Natl. Acad. Sci. U.S.A.* **83**, 7542–7546.

36 Popko, B., Puckett, C., Lai, E., Shine, H.D., Readhead, C., Takahashi, N., Hunt, S.W., III, Sidman, R.L., and Hood, L. (1987). *Cell* **48**, 713–721.

37 Privat, A., Jacque, C., Bourre, J.M., Dupouey, P., and Baumann, N. (1979). *Neurosci. Lett.* **12**, 107–112.

38 Roch, J.-M., Brown-Luedi, M., Cooper, B.J., and Matthieu, J.-M. (1986). *Mol. Brain Res.* **1**, 137–144.

39 Roach, A., Takahashi, N., Pravtcheva, D., Ruddle, F., and Hood, L. (1985). *Cell* **42**, 149–155.

40 Saxe, D.F., Takahashi, N., Hood, L., and Simon, M.I. (1985). *Cytogenet. Cell Genet.* **39**, 246–249.

41 Sidman, R.L., Conover, C.S., and Carson, J.H. (1985). *Cytogenet. Cell Genet.* **39**, 241–245.

42 Takahashi, N., Roach, A., Teplow, JD.B., Prusiner, S.B., and Hood, L. (1985). *Cell* **42**, 139–148.

43 Zeller, N.K., Hunkeler, M.J., Campagnoni, A.T., Sprague, J., and Lazzarini, R.A. (1984). *Proc. Natl. Acad. Sci. U.S.A.* **81**, 18–22.

NEURAL GROWTH

5

DEVELOPMENTAL SWITCHING OF GENES AND GENESIS OF THE VERTEBRATE CNS

SETSUYA FUJITA

Department of Pathology, Kyoto Prefectural University of Medicine, Kyoto 602, Japan

Analyses of birthdates and migration of neuroblasts in developing vertebrate embryos have revealed that there are highly regulated patterns in the genesis of neuronal and glial populations in the vertebrate central nervous system (CNS). At the beginning, there is a stage in which the neural tube is composed solely of matrix cells (Fig. 1, A, I). At this stage I, matrix cells proliferate only to multiply themselves (8). After several mitoses, matrix cells begin to produce neuroblasts (a term meaning immature neuron, composed of the Greek words "neuron" and "child"). This stage of neuron production is called stage II of cytogenesis of the CNS (9). There appears to exist a rigid and close correlation between the time-and-place of birth and the type of neuron differentiation determined at birth (19). As a general rule observed throughout the CNS, the neuroblasts produced first in respective regions mature into output type neurons which project longer axons out from the regions. The neuroblasts produced next usually mature into receptive type neurons, and those produced in the last phase become intercalating small local neurons. Detailed analysis in an individual region, however, reveals a highly individualized pattern of neurogenesis, although basically the pattern is within the frame of the general rule.

Once the neuroblast is differentiated from matrix cells, most fea-

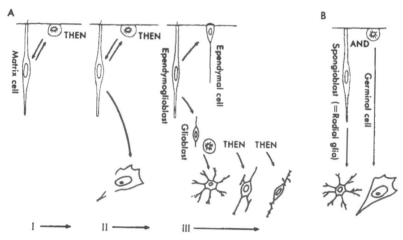

Fig. 1. Two theories of cytogenesis in the CNS. A: cytogenesis of the vertebrate CNS is divided into three consecutive stages (9, 10). In stage I, the neural tube is composed solely of matrix cells which perform an elevator movement (⇋). In stage II, matrix cells give rise to neuroblasts in preprogrammed order. When all the neurons are produced, matrix cells change into "ependymoglioblasts", common progenitors of ependyma and neuroglia. This is the beginning of stage III of cytogenesis. The ependymoglioblasts are rapidly differentiated into ependymal cells and glioblasts. The latter give rise to astrocytes, oligodendroglia, in sequence, and finally metamorphose themselves into the microglia (14). B: His' germinal cell theory and Rakic's radial glia. His distinguished two committed stem cells in the wall of the neural tube: rounded germinal cells and elongated spongioblasts. He believed that the former are specialized neuron-producing cells and the latter committed neuroglial precursors. Recently Rakic et al. (20) found that many elongated cells in the embryonic brain vesicles reacted strongly with anti-GFAP antisera and concluded, in confirmation of His' theory, that His' spongioblasts are nothing but the radial glia co-existing with neuron-producing cells. The sole evidence on which they depend, however, is the positive reaction (dots in Fig. 1B) of anti-GFAP anti-sera in the cells composing embryonic brain vesicles.

tures, if not all, of the future neuron are irreversibly fixed and cannot be altered by subsequent dislocation or environmental changes.

This is confirmed (4) in the mutant mice reeler in which location of the cortical neurons is drastically disordered but the type of neuron differentiation is unchanged. Transplantation into heterotopic sites (22) or explantation in vitro have both failed to change the original type of differentiation of neurons. It is likely that differentiation of the neuroblast is determined irreversibly when it is produced from the matrix cells at stage II of cytogenesis. Its determination takes place at an early G1 of the matrix cell (9).

When all the neurons are produced, stage II ends and the stage of neuroglia production begins (*14*). Matrix cells are now restricted to producing only non-neuronal cells and change into ependymoglioblasts, which are soon differentiated into ependymal cells and glioblasts. This is stage III of cytogenesis. The glioblast is a common progenitor of neuroglial cells of the CNS, and first differentiates into astrocytes, then oligodendroglia and finally metamorphoses into the microglia (*14*). Based on observations described above, cell differentiation in the developing vertebrate CNS can be explained with the simple scheme shown in Fig. 1A.

I. GLIAL FIBRILLARY ACIDIC PROTEIN (GFAP) AND MATRIX CELL

Recently, several authors (*1, 5, 20*) have reported that they observed a strong reaction of anti-GFAP in matrix cells at an early stage of development, and claimed that neuroglial differentiation proceeds in parallel with the production of the neuroblast (Fig. 1B). If matrix cells produce such a great amount of GFAP, their relationship to neuroglial differentiation would have to be re-examined.

Those investigators who have observed the positive GFAP reaction (*1, 5, 20*) have used antisera provided by Eng or by Bignami and Dahl. Curiously enough, none of these three has ever found any positive GFAP reaction in matrix cells at stages I and II with their own antisera. Critical examination is obviously necessary.

We have investigated this problem using chicken, mouse, rat, bovine, and human fetal brains and spinal cords by applying immunohistochemical staining, chemical analysis with SDS-polyacrylamide gel electrophoresis (SDS-PAGE), and immunoblotting (*14*), and came to a conclusion that GFAP is not present as protein in stage I and II matrix cells, at least, at a level detectable by present day techniques. This conclusion, however, does not appear definitive, since the possibility remains that the signal of the GFAP might be transcribed but not translated during stage II of cytogenesis.

II. GENE TRANSCRIPTION OF GFAP

In order to examine when and where the gene of GFAP is transcribed, we prepared cDNA from mRNA of fetal bovine spinal cord. The procedure is illustrated in Fig. 2. The specificity was checked by GFAP synthesis in a reticulocyte lysate system using hybridization-selected mRNA of the cDNA. The synthesized GFAP was identified by immunoblotting. First, this cDNA was nick-labeled with tritium and confirmed to hybridize with 2.4 kb mRNA of bovine brain (Fig. 3, third lane). Then, its cross-reactivity with rat brain mRNA was confirmed. A positive signal was found at 2.0 kb rat mRNA by Northern blotting (Fig. 3, second lane).

Occurrence of the GFAP-encoded mRNA was studied using this

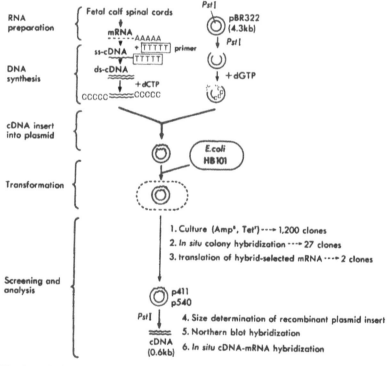

Fig. 2. Cloning procedure of cDNA for GFAP-encoded mRNA. Poly-(A)+ RNA is collected from fetal calf spinal cords. Complementary DNA was synthesized by reverse transcriptase and annealed with PstI-cut pBR322 plasmid. After transfection of Escherichia coli HB101, 1200 insert-positive clones were found. By colony hybridization with the calf spinal cord mRNA, 27 clones were selected and 2 of them were identified as GFAP-gene positive by synthesis of GFAP in reticulocyte lysate system. The two cDNAs were labeled with tritium and used as probes in the present study.

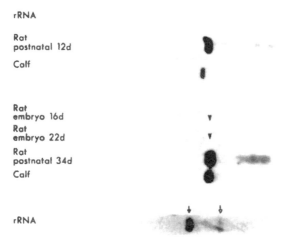

Fig. 3. Northern blot hybridization using GFAP-cDNA. Top lane is ribosomal RNA showing negative hybridization. The second lane is RNA from rat cerebral hemispheres at 12 days postnatal. Signal of GFAP-encoded mRNA is found on a band of 2.0 kb. The third lane, RNA from fetal calf spinal cord, shows positive hybridization at 2.4 kb. The fourth and fifth lanes are RNA from fetal rat cerebral hemispheres. Signals of GFAP-encoded mRNA are absent at the position (arrowheads) where it would appear postnatally. The seventh lane, RNA from rat cerebral hemisphere at 34 days postnatal; the eighth lane, RNA of fetal calf spinal cord. The last lane is rRNA of rat brain visualized with ethidium bromide to serve as molecular markers of 3.3 kb (⬇) and 1.7 kb (⇩).

cDNA probe. As shown in Fig. 3, the signal of GFAP-encoded mRNA was found only postnatally, and no trace of the signal was detected in the RNA fraction of fetal rat brains (embryonic day 16, 22). A similar observation was reported by Lewis and Cowan with mouse GFAP-mRNA (13). They first found a weak positive signal at postnatal day 2 which increased until day 9 followed by a slight decrease to the adult. In their experiment, embryonic mouse brains did not show any positive signal of GFAP-encoded mRNA. In chicken brains, Capetanaki et al. (1984) also obtained the same result (2), up to day 10 of incubation no trace of GFAP mRNA was detectable by Northern blotting but a strongly positive signal appeared in the spinal cord 2 weeks after hatching.

The result revealed that the occurrence of the GFAP-encoded mRNA parallels the appearance of the GFAP molecules in the developing CNS as described previously (13), and that, in rat as well as chicken and mouse brains, transcription of GFAP-encoded gene first occurs some time after stage III of cytogenesis commences, but is not concomitant with neuroblast production.

III. TIME-AND-PLACE OF mRNA OCCURRENCE STUDIED BY *IN SITU* HYBRIDIZATION TECHNIQUE

Applying Northern blotting, GFAP-encoded mRNA is found to appear first at stage III of cytogenesis but it is not clear in what cells it occurs and what space distribution it occupies. We studied this problem using *in situ* hybridization.

As shown in Fig. 4, the signal of GFAP-encoded mRNA was found exclusively on the cytoplasm of astrocytes.

Since we could confirm reliability of *in situ* hybridization, we applied the technique to study how GFAP-encoded mRNA is transcribed during development of the rat brain. At embryonic days 16 and 22 no positive cells were detectable in the cerebral hemisphere, corresponding to the result of immunohistochemistry of GFA protein and Northern blotting.

First, postnatally positive cells appeared in a subependymal posi-

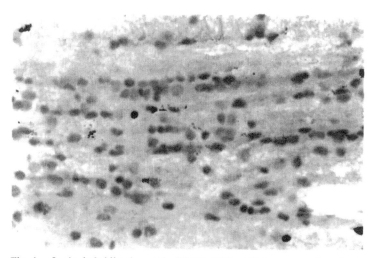

Fig. 4. *In situ* hybridization with GFAP-cDNA. White matter of rat brain stem. Cytoplasm of astrocytic cells shows positive signals of GFAP-encoded mRNA. None of the neurons or oligodendroglia as well as many astrocytes in the gray matter show any silver grains.

tion at day 2 and subsequently increased in number between days 12 and 23. These cells are immature neuroglial cells and may be identified as astrocytic cells.

IV. SWITCHING ON AND OFF OF THE GFAP-ENCODED GENE

Since it is widely believed that GFAP is a specific marker of the astrocyte, one may expect that all the astrocytes must be positive with GFA

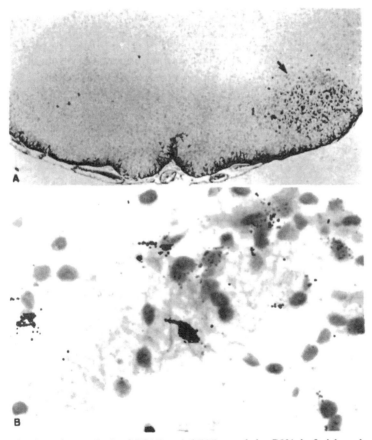

Fig. 5. Reactive synthesis of GFAP and GFAP-encoded mRNA in facial nucleus of the rat after transection of the facial nerve. A: 12 days after transection of the right facial nerve. Astrocytes in the injured nucleus (⬇) synthesize GFAP. The section is immunohistochemically stained with anti-GFAP antisera. B: 5 days after transection. Hybridization with GFAP-cDNA shows active synthesis of mRNA in and around the injured facial nucleus. Apparently GFAP synthesis is regulated by transcription of mRNA from the gene.

protein. This is not the case. In human as well as rat cerebral cortex, many astrocytes of layers II to VI are negative with immunohisto-chemical staining with anti-GFAP antisera. More striking is the bovine cortex; most astrocytes in the cortical gray are negative, in sharp contrast to strongly positive astrocytes in the subcortical white matter. The same is also true, to some extent, with gray matter of the brain stem. These astrocytes devoid of GFAP are all negative with GFAP-encoded mRNA. Figure 5 shows a cross section of rat brain stem at the level of the facial nucleus. The right facial nerve was cut 12 days before sacrifice. The section is stained with anti-GFAP antisera. Many astrocytes now strongly positive with GFAP appear in and around the right facial nucleus (arrow). In contrast, the left side which serves as normal control shows few astrocytes, if any, reacting positively with the antisera. Figure 5B is a result of *in situ* hybridization with GFAP cDNA 5 days after transection of the facial nerve. The positive signal is now detectable in many astrocytes. Immunohistochemical staining with anti-GFAP anti-sera also revealed positive reaction in these astrocytes. These observations indicate that transcription of the GFAP gene is reversible, and that the synthesis of GFAP is controlled directly by the transcriptional activity of this gene, as has been reported with synthesis of proteins of other intermediate filaments in various kinds of cells.

Summarizing these observations at protein and mRNA levels, we conclude that GFAP is not present in stage I or stage II of cytogenesis, and that the classical concept of "primitive spongioblasts" (*cf.* Fig. 1B) as committed precursors of neuroglia claimed to be present in an early phase of cytogenesis of the vertebrate CNS is not tenable.

V. DETERMINATION OF CELL DIFFERENTIATION AND ITS POSSIBLE GENETIC MECHANISM

It is now generally believed that development of the vertebrate, when viewed at the cellular level, proceeds by sequential steps in which potencies of progenitor cells become progressively and irreversibly restricted. This phenomenon is called "major differentiation" (*12*). The simplest hypothesis to explain the mechanism of the restriction of the potencies is to assume progressive and irreversible inactivations accumulating among functional subunits (*i.e.*, replicons) of chromosomal DNA (*11*).

It has been pointed out that the DNA portions that are irreversibly inactivated in the major differentiation are characterized by four extraordinary features (*11*): 1) incapability of RNA synthesis, 2) shortened and condensed in the interphase, 3) replicating late in the S-phase, and 4) this acquired feature of the inactivated DNA is inherited by the daughter cells and remains unchanged through subsequent mitoses.

VI. MAJOR DIFFERENTIATION OF MATRIX CELLS

In analyzing the elevator movement (*8*) by ^{3}H-thymidine autoradiography, cell cycle parameters have been measured in various species of animals (*13*). There is an unmistakable tendency of steady elongation of cell cycle and DNA synthetic times as the development proceeds. This tendency has been observed in the development of cell cycle change in all the animals so far studied. Matrix cells of neurectodermal origin, erythroblasts of chicken (*18*) endodermal cells in *Xenopus laevis* (*17*), ectodermal cells of cynops pyrrhogaster (*23*), cells of blastomeres of sea urchin embryo (*6*), *etc.* have been reported to show the same tendency.

According to the hypothesis of "major differentiation," the length of the S-phase is expected to become longer in differentiated cells in comparison with that of their undifferentiated precursors (*12*) as illustrated in Fig. 6.

At the beginning of the vertebrate ontogenesis, none of the replicons in the zygote are irreversibly inactivated, corresponding to the totipotent state in the major differentiation. DNA replicons in the cell begin to synthesize their DNA synchronously at the onset of the S-phase so that the overall rate of DNA synthesis shows a simple pulse-shaped curve (Fig. 6, at left). The length of the S-phase is expected to be short. When the cell progresses in steps of the major differentiation, *i.e.*, as the cell is differentiated, irreversibly inactivated replicons increase in number so that the S-phase becomes longer. The curve of the overall rate of DNA synthesis of the cell is now expected to have multiple peaks as shown in the right-hand diagram of Fig. 6.

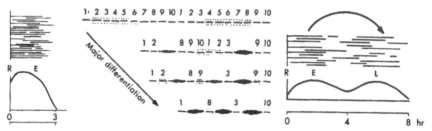

Fig. 6. Schematic drawing illustrating changes in chromosomes and rate of DNA synthesis in relation to cell differentiation (*11*). In the center, chromosomal changes during development are illustrated. In an undifferentiated cell (top), replicons composing the chromosome are not irreversibly inactivated, although only a few of them are actively transcribing mRNA (dotted segments). All replicons as shown in the figure at the left are early replicating (E) and rate of DNA synthesis (R) is expected to form a simple pulse-shaped curve. The absolute length of the S-phase should be short. While major differentiation proceeds, many replicons are irreversibly inactivated as shown in the condensed state in this diagram (center). They become late replicating (L) and make the curve of the rate of DNA synthesis complicated (right figures). The length of the S-phase also becomes longer with additional late replicating segments. The irreversible inactivation of genes is the genetic basis of determination of cell differentiation, and reversible on-off-switching of potentially active genes corresponds to functional modulation of the cell (*11*).

The tendency toward steady elongation of the S-phase in matrix cells of developing vertebrate embryos seems to support the notion that these cells steadily progress in the steps of major differentiation as the development proceeds.

VII. MAJOR DIFFERENTIATION AND FORMATION OF NEURONAL AND NEUROGLIAL CELLS

If we assume the above hypothesis (*3, 11, 12*) that irreversible inactivations of genes determine differentiation of the cell, and that the type of cell differentiation is determined by the combination of the irreversibly inactivated replicons, we can understand the characteristics of matrix cell differentiation and neuroblast production as follows.

Although matrix cells keep their epithelial morphology unchanged from the very beginning of neural plate formation to the end of stage II of cytogenesis, they change their state of (major) differentiation steadily as the development proceeds (*12*). When they repeat mitoses and enter into the G1 phase, irreversibly inactivated replicons increase in number and the cell accumulates steps of major differentiation. The

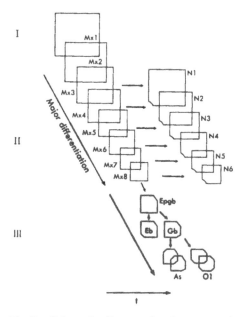

Fig. 7. Schematic diagram showing progression of irreversible differentiation (major differentiation) of matrix cells during development of the CNS (12). State of major differentiation of matrix cell population is steadily changing (from Mx1 to Mx8). From each state of major differentiation, specific neuroblasts (N1 to N6) are produced. Their specific states of differentiation are predetermined by those of their immediate precursors, the matrix cells. What determines the transition from matrix cells to neuroblast may be a common inactivation of one replicon that contains genes essential for DNA replication. I, II, and III in the figure correspond to stages of cytogenesis in the CNS and t represents time.

combinations of the inactivated replicons and their distribution patterns are supposed to be different in different cells; two daughter cells born from the same matrix cell inherit the same pattern of the inactivated replicons but can acquire new inactivations on different additional replicons forming different subclones in terms of major differentiation. Figure 7 shows one branch of such matrix cell subclones. Frames of Mx1 to Mx8 represent the magnitude of differentiation potencies of the matrix cells at respective stages of the major differentiation. When the major differentiation reaches a certain level, matrix cells can differentiate neuroblasts (commencement of stage II). Neuroblast differentiation in the vertebrate CNS is characterized by absolute repression of DNA replication; it is possible that neuroblast differentiation from a matrix cell may be determined by an irreversible repression of gene(s) directly or indirectly related to DNA replication. It is also possible that the differentiation of all neuroblasts (or neurons) is determined by one common additional inactivation of the replicon in the genome of the matrix cells at a certain state of major differentiation (Mx1~Mx8).

If one can assume this mechanism, it is easily understood why highly specialized neurons are produced at a given time and place during stage II of cytogenesis and why their future fate is irreversibly fixed at the time of birth of the neuroblasts.

Although this hypothesis of progressive gene inactivation was proposed more than 20 years ago (11, 12), it has been regarded as unlikely until quite recently. This is because of the widespread belief that the timing of replication of genes is fixed to the same pattern in all cells, differentiated and undifferentiated alike. In order to prove definitively that the timing of DNA replication of genes may change according to the state of the major differentiation, it was absolutely necessary to use a new technique of analysis that enabled us to determine the timing of DNA replication of the genes related to cell differentiation.

VIII. ANALYSIS OF TIMING OF GENE REPLICATION USING SPECIFIC cDNA PROBES FOR THE GENE

In 1981, Furst et al. (15) and Epner et al. (7) introduced a technique to determine the timing of gene replication in the S-phase (15). They first synchronized cultured cells at the mitotic phase using colchicine or vinblastine and flash-labeled the cells at early S-phase or at late S with bromodeoxyuridine (BrdU). After extracting DNA from the cells and treating it with restriction enzymes, they centrifuged the DNA fragments in CsCl density gradient to separate heavy DNA labeled with BrdU that was synthesized at the early (or late S) and the light DNA free from BrdU-label that was not under synthesis during that period. Thus, early or late replicating DNAs could be separated. Two DNA fractions were then run on agarose electrophoresis, transblotted to nitrocellulose membrane and hybridized with cDNA probe labeled with ^{32}P. If the gene corresponding to the cDNA is early replicating, for example, the positive signal is detectable on the early replicating

DNA fraction. This technique has been successfully applied to a number of genes, including those related to glycolysis and nucleic acid metabolism that are essential for survival of cells. These genes are called "housekeeping genes" and are always found to be early replicating. In contrast, genes related to enzymes or proteins that are functional only in some specialized cells are found to be either late or early replicating. These differentiation-related genes are early replicating in the cells where these genes are actually expressed, but many of them are found to be late replicating in those cells that do not utilize these genes; β-globin genes are early replicating in erythroid MEL cells but late replicating in HeLa cells of squamous epithelium origin or in plasma cells. Goldman *et al.* (*16*) added other examples such as genes of liver enzymes, α_1-antichymotrypsin, α_1-antitrypsin, phenylalanine hydroxylase that are late replicating in HeLa cells, and milk protein β-casein that is also found late replicating in non-mammary cells. Due to the technical difficulty of synchronizing the cells, these data are restricted only to cell strains *in vitro*. No application has been made with cells of the CNS. Recently, however, we have found that the replication timing analysis of a specific gene can be applied to cell systems *in vivo* and *in*

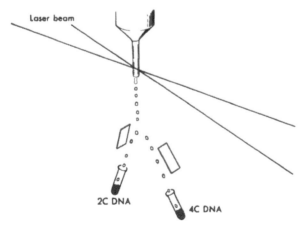

Fig. 8. Separation of early and late S cells by a cell sorter. If BrdU-labeled cells are separated according to their nuclear DNA content, early S nuclei (2.5–2C) and late S nuclei (3.5–4C) can be obtained separately. After cutting DNA by restriction enzymes, the DNA fragments can be analyzed by Southern blotting to identify the timing of replication of a gene.

vitro without synchronizing a cell population. This technique uses a cell sorter to separate BrdU-labeled cells into early and late S cells according to their DNA content. Cells with 2 ($\sim$2.5) C DNA are regarded as early S cells and (3.5$\sim$) 4C cells are late S cells (Fig. 8). Investigations are now under way using this technique.

If we can assume that differentiation of neurons and neuroglia is determined by the irreversible repression of replicons during neurogenesis, many important problems of cell differentiation in the CNS such as transition of stage II of cytogenesis into stage III can be explained in a simple way. Namely, if major differentiation of matrix cells progresses and neuron-essential gene(s) such as those for excitable membrane, receptors or transmitters, *etc.* are irreversibly inactivated in a matrix cell, it can no longer produce neurons. What it can differentiate is only non-neuronal cells, *i.e.*, neuroglial cells. Phenotypically, this inactivation must create the beginning of stage III of cytogenesis.

The progressive gene inactivation hypothesis or the major differentiation hypothesis can explain nicely the sequential occurrence of stages I, II, and III and production of specific cells in the development of the vertebrate CNS. Not only can it explain the cytogenesis but it also enables analysis of the genetic mechanism of cellular differentiation in the developing CNS in molecular terms; when, where, and what kind of specific gene or genes is inactivated to determine the differentiation of various kinds of neurons or neuroglia can be mapped by this technique. An extremely complicated pattern of cell differentiation in the CNS, though so it appears at first glance, might turn out to be the result of simple hierarchical repressions of certain classes of essential genes.

IX. IRREVERSIBLE GENE INACTIVATION AND NEURAL PLASTICITY

We have proposed a hypothesis that cellular differentiation of the CNS is realized by progressive and irreversible inactivations of replicons accumulating during ontogeny. According to this hypothesis, cells in the CNS are characterized by very rigid cytodifferentiation. Paradoxically, it is well known that the vertebrate CNS, particularly of the primate, is highly plastic at least in functional aspects. How is this neural plasticity related to the irreversible differentiation of the cells of the CNS?

According to the hypothesis, the major differentiation of a cell is determined by irreversible inactivation of *some* replicons. The inactivated replicons are supposed to be a minority in number among all the replicons of the genome. The rest, the majority of replicons are in a potentially active state, and can be switched on and off in response to the signals of the intra- and extracellular environment. This type of reversible expression of cellular differentiation which has been called "minor differentiation" (*11*) is characterized by the reversible synthesis of mRNA and proteins. For analysis of the minor differentiation, Northern bloting and dot blot hybridization can be applied at histological or cell-populational levels. At the cellular level, the technique of *in situ* hybridization is particularly useful.

In parallel with these analyses on nucleic acids, the occurrence of specific proteins can be analyzed by applying methods of immunohistological staining. Enzyme-linked immunosorbent assay (ELISA), SDS-polyacrylamide gel electrophoresis (SDS-PAGE) are available to identify and quantify translational products of the mRNA. Key substance(s) to perform this kind of molecular analysis are specific probes for the gene and its mRNA (cDNA) or for its translational products (antibodies). As is well known, recent progress in molecular biology now offers us the possibility of obtaining any specific probes, if desired, either in the form of cDNA to nucleic acids or antibodies to proteins.

SUMMARY AND PROSPECTS

Development of the vertebrate, viewed on a cellular level, proceeds by sequential steps in which potencies of progenitor cells become progressively and irreversibly restricted. This type of irreversible differentiation of a cell is called "major differentiation." The simplest hypothesis to explain the genetic mechanism that realizes this phenomenon is to assume progressive and irreversible inactivations accumulating among functional subunits (replicons) of chromosomal DNA. It has been pointed out that the DNA portions that are irreversibly inactivated become incapable of RNA synthesis, are shortened and condensed in the interphase, and replicate late in the S-phase. This acquired feature of the inactivated DNA is inherited by the daughter cells and kept unchanged through subsequent mitoses. Although it is some 20 years ago that this hypothesis was proposed, only recently has it become possible to test it by using cDNA probes of specific genes and identifying the timing of their replication in the S-phase of various types of cells at differentiated and undifferentiated states. The results of the experiments so far reported provide strong evidence to support this hypothesis of major differentiation.

In the development of the vertebrate CNS, the period of cytogenesis is divided into three consecutive stages, I, II, and III. The sequential nature of the differentiative behavior of matrix cells can now be explained by the hypothesis of progressive gene inactivations that accumulate in genomes of matrix cells during the development, as follows. Different types of neurons are produced from matrix cells at different states of *major differentiation*. Irreversible inactivation of genes progresses during stage II of cytogenesis and when genes essential for neuronal differentiation (such as one or several structural genes for synthesis and maintenance of excitable membranes, receptors or transmitters, *etc.*) are inactivated in matrix cell-genome, the cell can no longer differentiate neurons but necessarily produces only non-neuronal elements, *i.e.*, neuroglia and ependymal cells. Thus the consecutive occurrence of stages I, II, and III can be understood.

Recently, however, a few investigators have claimed that differentiation of neurons and neuroglia may not be sequential, thereby precluding the simple explanation of neurogenesis. A crucial point is the histochemical reaction of anti-GFAP antisera.

We therefore re-examined immunohistological and immunochemical reactions of cytoskeletal proteins of the developing CNS, with special reference to GFAP in the developing chicken, murine, rat, bovine, and human brains and spinal cords. Matrix cells in the neural tube of all the materials examined were stained positive for vimentin and tubulin, but negative for GFAP in stages I and II of cytogenesis. Immunoblotting of the same materials confirmed that GFAP antigen is not present in the matrix cells beyond a detectable level during stages I and II of cytogenesis. It was confirmed that GFAP first becomes detectable at stage III.

There is still a possibility that the GFAP gene is active during stages I and II, producing mRNA but no protein. We studied, in the present investigation, the occurrence of GFAP-encoded mRNA by Northern blotting and *in situ* hybridization. It became clear that the gene for GFAP is repressed during stages I and II of cytogenesis while neurons are being produced, and only at stage III is the switch of the GFAP gene turned on. Gene expression of GFAP in astroglia and ependymal cells is reversible but expressible only in stage III.

There are two types of gene switching in the development of the vertebrate CNS. One is progressive inactivation of genes that is an irreversible switch and determines the differentiation of cells. Combination of the inactivated genes in this type of irreversible switching characterizes the type of cell differentiation realizable after maturation.

On the other hand, the second type of gene switching is completely reversible and may respond readily to the environmental signals. It is this latter type of gene switching that is important in giving plasticity to the cells of the CNS. In contrast, the former type of irreversible switching mechanism confers on the cells of CNS the stable genetic framework of neurogenesis that has been inherited over tens of millions of years with very slow evolutionary modification, giving to the species their brains with specific morphology and behavioral identities of their own.

In future research on the genetic mechanism of brain morphogenesis and function, we should take the presence of these two mechanisms of switching into consideration.

Fortunately, techniques to analyze these two mechanisms are already in the hands of present-day neurobiologists. We can expect immense progress in the understanding of molecular aspects of the developing vertebrate CNS in the next 10 years.

REFERENCES

1 Antanitus, D.S., Choi, B.H., and Lapham, L.W. (1976). *Brain Res.* 103, 613–616.
2 Capetanaki, I.G., Ngai, J., and Lazarides, E. (1984). In *Molecular Biology of the Cytoskeleton*, eds. Borisy, G.G., Cleveland, D.W., and Murphy, D.B., pp. 415–434. New York: Cold Spring Harbor Lab.
3 Caplan, A.I. and Ordahl, C.P. (1978). *Science* 201, 120–130.
4 Cavines, V.S., Jr. (1982). *Dev. Brain Res.* 4, 293–302.
5 Choi, B.H. (1986). *J. Neuropathol. Exp. Neurol.* 45, 408–418.
6 Dan, K.S., Tanaka, K., Yamazaki, K., and Kato, Y. (1980). *Dev. Growth Differ.* 22, 589–

598.

7 Epner, E., Rifkind, R.A., and Marks, P.A. (1981). *Proc. Natl. Acad. Sci. U.S.A.* **78**, 3058–3062.

8 Fujita, S. (1962). *Exp. Cell Res.* **28**, 52–60.

9 Fujita, S. (1964). *J. Comp. Neurol.* **122**, 311–328.

10 Fujita, S. (1965). *J. Comp. Neurol.* **124**, 51–59.

11 Fujita, S. (1965). *Nature* **206**, 742–744.

12 Fujita, S. (1975). *Symp. Cell Chem.* **27**, 97–105.

13 Fujita, S. (1986). *Curr. Top. Dev. Biol.* **20**, 223–242.

14 Fujita, S., Tsuchihashi, Y., and Kitamura, T. (1981). In *Glial and Neuronal Cell Biology*, eds. Vidrio, E.A. and Fedroff, S., pp. 141–169. New York: Alan Liss.

15 Furst, A., Brown, E.H., Braunstein, J.D., and Schildkraut, C.L. (1981). *Proc. Natl. Acad. Sci. U.S.A.* **78**, 1023–1027.

16 Goldman, M.A., Holmquist, G.P., Gray, M.C., Caston, L.A., and Nag, A. (1984). *Science* **224**, 686–692.

17 Graham, C.F. and Morgan, R.W. (1966). *Dev. Biol.* **14**, 439–460.

18 Holtzer, H., Weintraub, H., Mayne, R., and Mochan, B. (1977). *Curr. Top. Dev. Biol.* **7**, 251–265.

19 Jacobson, M. (1970). *Developmental Neurobiology.* New York: Holt, Reinhardt & Winston.

20 Levitt, P., Cooper, M.L., and Rakic, P. (1983) *Dev. Biol.* **96**, 472–484.

21 Lewis, S.A. and Cowan, N.J. (1985). *J. Neurochem.* **45**, 913–939.

22 Nakamura, H., Nakano, K.E., Igawa, H.H., Takagi, S., and Fujisawa, H. (1986). *Cell Differ.* **19**, 187–193.

23 Yamamoto, K.Y., Yamazaki, K., and Kato, Y. (1980). *Dev. Growth Differ.* **22**, 79–82.

6

EARLY CELL COMMITMENT AND ENVIRON-MENTAL FACTORS IN THE DEVELOPMENT OF PERIPHERAL NERVOUS SYSTEM IN BIRDS

NICOLE M. LE DOUARIN AND ZHI GANG XUE

Institut d'Embryologie du CNRS et du Collège de France, 94736 Nogent-sur Marne Cedex, France

Previous studies essentially based on the use of the quail-chick marker system reported in detail the precise level of origin of the various cellular components of the peripheral nervous system (PNS) along the neuraxis in the avian embryo (see *11*). Similarly, the respective contributions of the placodal ectoderm and of the neural crest to the cranial nerve sensory ganglia could be established by virtue of exchanges of embryonic territories between embryos of the quail and chick species. This led to the construction of a fate map of the presumptive PNS territories on the neural ectoderm at the late neurula stage (*i.e.*, in the embryo at 2–3 days of incubation—E2.3—from the time the neural crest cells start to migrate in the head at the 8-9-somite-stage to the complete closure of the neural tube) (see *11, 13, 14* for review and *5*). Figure 1A shows a schematic representation of this map, in which it can be seen that a regionalisation exists in the neural crest, each region yielding a particular set of peripheral ganglion precursor cells.

However, it also appeared that such a regionalisation did not correspond to strict restriction in crest cell developmental potentials (Fig. 1B). This was shown by performing heterotropic transplantations

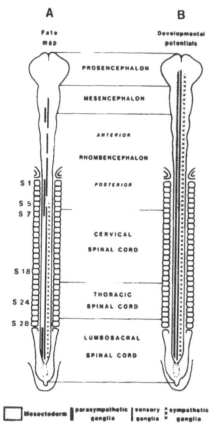

Fig. 1. A: fate map of the presumptive territories along the neural crest yielding the mesectoderm, the sensory, parasympathetic, and sympathetic ganglia in normal development. B: developmental potentials for the same cell types shown in the fate map are indicated. In fact, if neural crest cells from any level of the axis are implanted into the appropriate site of a host embryo, they can give rise to almost all the cell types forming the various kinds of PNS ganglia. This is not true, however, for the ectomesenchymal cells (also called mesectoderm), whose precursors are confined to the cephalic area of the crest down to the level of somite 5. S, somite.

of neural primordia between quail and chick embryos (these were also heterochronic due to the temporal cranio-caudal gradient of neural crest development); such transplantations led neural crest cells from a given region of the embryo (*e.g.*, the head) to develop in another (*e.g.*, the trunk). In fact, when transplanted into a different embryonic territory, neural crest cells were capable of migrating along the pathways and settling in the target tissues corresponding to their new site. Moreover, they differentiated into neuronal cell types characteristic of these target tissues rather than into the neuronal types they would normally have yielded if left *in situ*. This means i) that the neural crest developmental potentials are broader than the phenotypes they actually express during development, and ii) that the neural crest displays during ontogeny a considerable amount of plasticity which enables this embryonic structure to adjust to external cues.

Potentialities to express virtually all the phenotypes encountered in PNS ganglia are present along the whole neuraxis; this fact implies that the target organ exerts a selection between these possible phenotypes.

The next problem concerns the mechanisms through which the embryonic microenvironment of neural crest cells during and/or after their migration influence their differentiation.

Experiments have been devised, *in vivo* and in tissue culture, to try and define the state of commitment of the crest cells when they start migrating and later on aggregate to form the various PNS ganglia. They have led to the notion that, early in neural crest development, distinct neuronal cell lineages become segregated and are characterised by different requirements for their further differentiation.

Some of these experiments will be briefly described in this article.

I. BACK-TRANSPLANTATIONS OF QUAIL PERIPHERAL GANGLIA INTO THE NEURAL CREST MIGRATION PATHWAY OF E2 CHICK EMBRYOS

The rationale behind these experiments was to subject developing PNS ganglia to the microenvironment of the dorsal embryonic structures (involving the neural tube, the notochord and the somites) at the time when they normally influence neural crest cells and lead them to differentiate into sensory and sympathetic ganglia.

Fragments of ganglia, containing about 2,000 cells, were introduced into a slit made between somites and neural tube and the host embryo was allowed to develop for one to 8 days post grafting. Then it was sacrificed and the cells derived from the transplant were identified through the quail nuclear marker (Fig. 2). Surprisingly, the

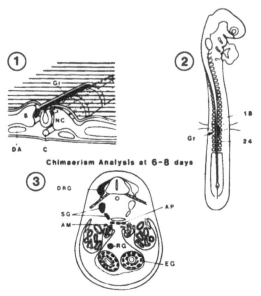

Fig. 2. Diagram showing the experimental procedure followed in the back-transplantation of quail PNS ganglia into the chick neural crest migration pathway. 1 and 2: the positioning of the graft. 3: the various host crest derivatives in which quail cells are found in the 6–8 day chick host. AM, adrenal medulla; AP, aortic plexus; C, notochord; DA, dorsal aorta; DRG, dorsal root ganglion; EG, enteric ganglia; Gr, graft; NC, neural crest; RG, ganglion of Remak; S, somite; SG, sympathetic ganglia.

grafted tissue loses its cohesiveness when grafted in this embryonic area, while it does not if grafted in other regions of the body at the same age. As early as 24 hr post-grafting, its component cells become dispersed in the host somitic area. After 48 hr, the dispersed cells home to the host's neural crest-derived tissues.

Although the phase of dispersion varies little, homing of the grafted cells into host tissues changes with the nature—sensory or autonomic —and the age of the grafted ganglia. Quail cells populating the host dorsal root ganglia (DRG) and differentiating there into sensory neurons and glia were found only after grafts of DRG and of proximal cranial sensory ganglia (e.g., superior-jugular ganglionic complex of nerves IX and X). Autonomic (ciliary, Remak, sympathetic) ganglia, when transplanted under similar conditions, never gave rise to sensory neurons in the host DRG. Only rarely were a few glial cells of graft origin found in this position (7, 10, 15; see also 11, 13, 14 for reviews).

In contrast, Schwann cells and sympathetic autonomic derivatives (*i.e.*, sympathetic neurons, chromaffin cells) were obtained after grafts both of DRG and autonomic ganglia. It is interesting to notice that back-transplantation of ganglia originating from cephalic neural crest whether sensory (jugular-superior, nodose, petrosal) or parasympathetic (ciliary) provided the host's gut with ganglionic cells, while the ganglia arising from the truncal neural crest (DRG, sympathetic ganglia) did not.

Moreover, the capacity of quail DRG cells to populate the host sensory ganglia was found to be developmentally restricted to the period during which sensory neuroblasts are still cycling. As shown by Schweizer *et al.* (*19*), if DRG were taken for implantation at a time when all the sensory neurons had withdrawn from the cell cycle (an event which occurs first in the lateroventral and then in the mediodorsal regions of the ganglion and which is completed at E8 in the quail), the transplants gave rise exclusively to ganglion cells of the sympathetic autonomic type. This suggested that post-mitotic neurons did not survive when grafted into a younger host embryo.

It has been documented, mainly in tissue culture experiments, that survival and neurite regeneration of peripheral neurons are dependent upon growth factors, the best example of which is nerve growth factor (NGF). It is conceivable that in the 2-3-day old embryo such growth factors are either not yet produced or do not attain a concentration sufficient to ensure neurite outgrowth and maintenance of the grafted neurons.

Dupin (*6*) analysed in detail the multiplication of quail ciliary ganglion cells after back-implantation into the chick crest cell migration pathway. She found that particularly large numbers of quail cells incorporated [^{3}H]thymidine during the dispersion phase and also, to a lesser extent, after migrating to the neural crest derivatives of the host. The expansion of the ganglion cell population was much greater with younger ganglia and multiplication was higher in the graft than *in situ* over the same period of time. Since she also showed that post-mitotic neurons die after transplantation, one has to conclude that the cells which divide in the host must belong to the non-neuronal cell population of the ganglion. Moreover, proliferation factors, to which the non-neuronal cells of the grafted ganglia are able to respond, must be present in the 2–4-day embryonic axial structures.

The fact that the post-mitotic neurons do not survive after grafting into a younger chick was also demonstrated in the case of nodose ganglion transplants. Due to the mixed placodal and crest origin of the distal ganglia of certain sensory nerves (geniculate ganglion of nerve VII; petrosal of nerve IX and nodose of nerve X), it is possible to label selectively their neuronal or non-neuronal (destined to become glial) component cells. One can then follow selectively the fate of these cell categories in the back-transplantation system (1, 7). Non-neuronal cells of the nodose and petrosal ganglia are labelled when a chick embryo is used as a host for the isotopic and isochronic implantation of a quail neural primordium at the rhombencephalic level. The chick epibranchial placodes then provide the ganglia with neuronal cells, while the quail crest cells give rise to the entire non-neuronal cell population. If the host is a quail and the donor of the rhombencephalon a chick, the reverse situation will prevail.

Back-transplantation into a chick at the adrenomedullary level of a nodose ganglion with quail neurons and chick non-neuronal cells resulted in no colonization of the host crest derivatives by graft-derived cells. Only a few quail fibroblasts could be recognised at the graft site. In contrast, quail cells massively invaded the host when the grafted chimaeric nodose ganglion contained a quail non-neuronal cell population: numerous quail neurons and glia were found, not only in adrenal medulla, sympathetic ganglia and plexuses but also in enteric ganglia. However, no quail-labelled neurons and only very few glial cells were found in the DRG of the host following the graft either of a nodose or a petrosal ganglion with quail-labelled non-neuronal cells (1). Recently Fontaine-Pérus et al. (7) implanted petrosal and nodose ganglia whose non-neuronal cells carried the quail marker into the vagal crest migration pathway. They demonstrated the ability of these labelled cells to differentiate, not only into enteric ganglia (as expected), but also into type I and II carotid body cells.

1. A Cell Line Segregation Model

To account for these experimental results, a model was proposed in which two cell lines of peripheral ganglion progenitor cells become segregated early during neural crest ontogeny (12–14) (Fig. 3). Devel-

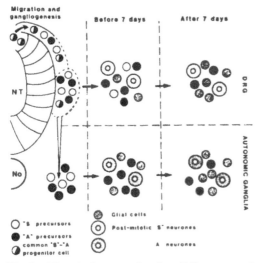

Fig. 3. Hypothesis accounting for cell line segregation during PNS ontogeny. Two types of precursors, sensory ("S") and autonomic ("A"), arise from a common progenitor during neural crest cell individualisation and/or migration and in the early steps of gangliogenesis.

opmental potentials of the two postulated precursors are already restricted to either the autonomic or sensory pathways in the migrating crest or very shortly after they have stopped migrating. Developing peripheral ganglia should thus contain a mixed population of both autonomic and sensory precursors. The fact that no sensory neurons can be obtained in the back-transplantation experiments from either distal sensory or autonomic ganglion grafts, while autonomic derivatives arise from all types of grafted peripheral embryonic ganglia may reflect different survival requirements of the two types of precursors; the sensory precursors are apparently able to survive and differentiate only in ganglia situated in close proximity to the central nervous system (CNS), while autonomic precursors must persist in all types of PNS ganglia. Moreover, since sensory neurons of graft origin do not arise in the host when the transplanted DRG is removed from quail embryos older than E7, it can be deduced that neuroblasts of the sensory type disappear from sensory ganglia themselves after this stage. As this corresponds to the time when all the DRG neurons of the quail have withdrawn from the cell cycle (19), it follows that sensory neuron

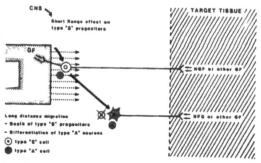

Fig. 4. Diagram showing the different survival requirements of the sensory (type "S") and autonomic (type "A") progenitor cells of the PNS at the time of early gangliogenesis. Type S progenitors have to be provided with a GF from the CNS to survive and extend neurites. It is only later that they find in the target organ NGF and/or other GF able to fulfil their needs. Type A progenitors can survive and differentiate at long distance from the CNS. In sensory ganglia, they survive for long periods of time but neither extend neurites nor differentiate. Type S precursors do not survive if they happen to migrate to ganglia distant from the CNS.

precursors either become post-mitotic neurons or disappear from the DRG after E7 in quail embryos. In contrast, autonomic precursors remain in all types of peripheral ganglia, but their proliferation and differentiation can be triggered only by an appropriate stimulus—such as may be found in the microenvironment of a younger host in the back-transplantation experiment system.

In support of this view is the fact that all the sensory ganglia whose neurons originate from the neural crest develop in contact with the CNS. This is true not only for the DRG but also for the proximal part of the trigeminal ganglion (situated on cranial nerve V) and for the superior jugular complex corresponding to the root ganglia of nerves IX and X. It seems therefore that the CNS exerts a short-range positive effect on the survival, and perhaps also on the differentiation, of precursor cells of the sensory lineage (Fig. 4).

2. Does There Exist a Particular Cell Lineage from Which the Enteric Neurons Are Derived?

In the cell line segregation model presented in the 1984 and 1986 articles *(12–14)* the term autonomic was used to designate all the peripheral neurons which did not belong to the sensory type. The latter can be defined by its function, which is to transfer sensory inputs from the periphery to the CNS through a centripetal axonal process joining the sensory ganglia to either the spinal cord or the brain. The term "autonomic" therefore groups together ganglionic neurons belonging to the sympathetic, parasympathetic, and enteric nervous systems.

In fact, the criteria used in all the back-transplantation experiments to define the nature of a given neuron were its localisation in the host (e.g., in a sensory, a sympathetic or an enteric ganglion) and the neurotransmitter or neuropeptides it contained.

In the first back-transplantation experiments that we did in 1978 (10) we had already noticed that quail cells migrated into the gut when ciliary, but not Remak, ganglia were back-transplanted. Later, we discovered that ciliary ganglion grafts were able to colonise the digestive tract only if taken from the donor before embryonic day 7. Therefore, the capacity to provide the host with cells invading the gut appears restricted both by the nature of the ganglion and the developmental stage of the graft. We realised recently that the relevant characteristic, as far as the nature of the ganglion is concerned, pertains to the level of the neuraxis from which it originates. For example, DRG, even taken from E5 quails and grafted at the vagal level where the neural crest cell migration pathway to the gut is localised, do not provide the host's intestine with graft-derived cells, while all kinds of cephalic ganglia (nodose, petrosal, jugular superior, ciliary) do so. This means that in fact this assay allows two distinct categories of "autonomic" precursors to be distinguished: *"autonomic sympathetic,"* with adrenergic differentiation capabilities, and *"autonomic enteric,"* which develop in the gut.

Although these two types of precursors exist throughout the neural crest, as attested by heterotopic transplantations and other types of experiments (see *11, 13, 14,* and *16*), their distribution along the cephalocaudal axis is uneven. A prevalence of the "autonomic enteric" over the "autonomic sympathetic" precursors exists in the cephalic-vagal crest while the inverse is true in the cervicotruncal region.

As for sensory neuronal precursors, the survival time of the enteric precursors in sites outside the gut is short. This can be deduced from the fact that they are not detectable by the back-transplantation assay in quail ciliary and dorsal root ganglia when the latter are removed from E7 and E5 embryos, respectively.

According to this view, the neural crest is a mixture of cells in which the developmental potentialities become progressively restricted to different phenotypes. There is no reason so far to think that such a precommitment affects (at least as far as the PNS is concerned) their migratory behaviour. It would therefore follow that migration and gangliogenesis involve a heterogeneous population of cells with different degrees of commitment, whence the contention that development of a particular cell type in a given embryonic site depends on the presence of survival and growth factors acting selectively on a specific set of precursor cells. To further document this question, it is necessary to: i) identify and, if possible, isolate each of the crest cell precursors so defined, and ii) determine what growth factor(s) they are able to respond to.

The next experimental approaches aimed at answering these questions concern the *in vitro* identification of the "sympathetic autonomic" precursors present in DRG on the one hand and the influence of the CNS on the differentiation of sensory ganglia on the other hand.

II. CHARACTERISATION OF THE SYMPATHETIC-AUTONOMIC PRECURSORS *IN VITRO*

The presence in sensory ganglia of resting autonomic precursors has been confirmed in culture experiments (*20, 21*). Adrenergic expression was used as an indicator of autonomic differentiation, and advantage was taken of the *in vitro* system to characterise the phenotype more fully, to identify the precursors capable of developing along this pathway and to define the conditions required for them to do so.

DRG were dissected out of quail embryos taken between 10 to 15 days of incubation, ages at which all the sensory neurons are unquestionably post-mitotic (*4, 19*), dissociated to single cells and plated on collagen-coated coverslips where they were grown at a density of 7×10^4 cells/dish in conventional tissue culture medium (MEM supplemented by horse serum, NGF and 11-day chick embryo extract). Under these conditions, the sensory neurons rapidly regenerated their neurites and the non-neuronal cells grew profusely, approaching confluence within 5–6 days (Fig. 5a).

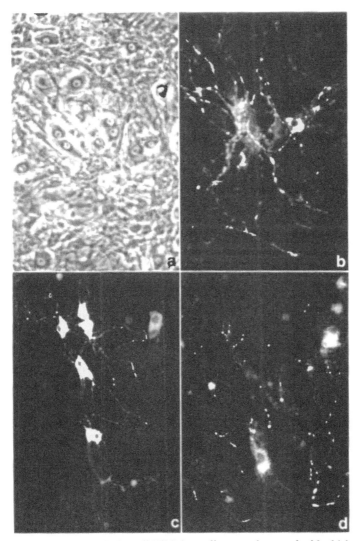

Fig. 5. Culture of E10 quail DRG in medium supplemented with chick embryo extract. a: phase contrast of a 6-day culture showing neurons and non-neuronal cells at confluence. $\wedge$: 196. b: culture (7 days) treated with glyoxylic acid to evidence CA content. Note the numerous fibers with strongly fluorescent varicosities and cell bodies. $\times$: 297. c, d: five-day culture treated immunocytochemically for tyrosine hydroxylase (c) and substance P detection (d). It is clear that these antigens are present in different cell populations. $\times$: 294.

At daily intervals after plating, the cultures were treated for immunocytochemistry with an antibody to tyrosine hydroxylase (TH). The appearance of immunoreactive cells was found to be highly dependent on the presence of chick (or quail) embryo extract in the medium. In its absence, although neuronal survival was apparently unaffected (provided that NGF was maintained), non-neuronal cell proliferation was greatly impaired and there was a 96–100% reduction in the number of TH-positive cells that ultimately differentiated. When the medium contained optimal concentrations of embryo extract, a few cells with TH immunoreactivity appeared after 3–4 days of culture. None was ever found before the end of the third day of culture. Their number then rose rapidly, reaching a maximum between 6 and 9 days, when they represented approximately 2% of the number of ganglion cells initially plated. The majority were small, rounded and multipolar with long, branched processes whose varicosities were particularly immunoreactive (Fig. 5c).

Glyoxylic acid treatment of cultures of 4 days and onward revealed intracellular catecholamines (CA), particularly at the level of the fibre network (Fig. 5b). Furthermore, dense-core vesicles typical of CA storage organelles were observed by electron microscopy of permanganate-fixed sections of cultures. Finally, the presence of CA-synthesising cells was confirmed biochemically by measuring the conversion of exogenous tyrosine to CA. Undetectable in cultures that had been grown for less than 4 days, appreciable amounts of radioactive noradrenaline were found in older cultures after a 4 hr incubation with [^{3}H]tyrosine. This finding implied that, in addition to TH, dopa decarboxylase and dopamine-β-hydroxylase were active in the cultures. The increase in the ability to produce CA that occurred between days 4 and 9 paralleled the increase in the number of TH-immunoreactive cells over the same period. Results of recent experiments using the technique of *in situ* hybridisation confirm that mRNA coding for TH appears after 3–4 days in DRG cultures, whereas it is absent from DRG *in situ* (Xue *et al.*, in preparation).

In addition to their capacity to synthesise and store CA, authentic adrenergic neurons also have the ability to accumulate extracellular CA *via* a high affinity transport system (*8*). The TH-positive cells developing in DRG ganglion cultures were found, in their overwhelming majority, to possess an uptake system of this sort (Xue *et al.*, in preparation). In consequence, it could be assumed that the TH antibody was identifying cells that were truly adrenergic as judged by several varied criteria.

The autonomic/sensory cell line segregation hypothesis predicts that the adrenergic cells differentiate from precursors that are distinct from post-mitotic neurons of the DRG. Besides the fact that most of the TH-positive cells were very dissimilar in size and shape to recognisable primary sensory neurons in the cultures, this contention was supported by several lines of evidence. Immunocytochemical investigations using a monoclonal antibody to substance P revealed that TH- and substance P-positive cells constituted two entirely non-overlapping populations; no coincidence of the two markers was ever found, either in cell bodies or fibres, in cultures of any age (Fig. 5c, d). Similarly, none of the cells capable of taking up CA specifically was ever labelled by the substance P antibody (22).

More compelling evidence for the non-neuronal origin of the adrenergic cells in the DRG cultures was obtained when it was shown that they and/or their precursors were able to proliferate. Exposure to [^{3}H]thymidine for various lengths of time resulted in the labelling of TH-positive cells at all stages of culture. In particular, it was shown that approximately half of the first TH-positive cells to appear (at 4 days of culture) had incorporated the radioactive label at some time during the previous 96 hr, *i.e.*, they had divided before differentiating. Evidence was also obtained that, as in the case of sympathetic neuron precursors *in vivo* (18), cells displaying the adrenergic phenotype could themselves divide. In contrast, substance P-containing neurons were never labelled after exposure to [^{3}H]thymidine, irrespective of the time at which the radioactive precursor was added; their post-mitotic status was thus confirmed.

Finally, an interesting observation suggested that the adrenergic precursors are recognisable in the DRG cell population before being put into culture. Cells, morphologically indistinguishable from those that subsequently become TH-positive, were found to take up [^{3}H] noradrenaline by a specific, desmethylimipramine-sensitive process, not only in young (12–48 hr) cultures, but also in freshly removed E10 DRG. Their non-neuronal nature was strongly suggested by their morphology; furthermore, none of the cells displaying uptake properties in short-term cultures expressed substance P or neurofilament protein immunoreactivity (Xue *et al.*, in preparation). CA uptake, exclusively by a non-neuronal cell population, has also been reported in 24 hr cultures of dissociated chick DRG (*17*).

In consequence, it seems clear that a population of non-neuronal cells with autonomic sympathetic potentialities is present in avian DRG, where it can be identified by its ability to take up CA. It is therefore not totally undifferentiated with respect to the adrenergic phenotype, but full expression requires exposure of the cells to appropriate environmental conditions, which *in vitro* are provided to them by factors contained in embryo extract.

III. EXPERIMENTAL EVIDENCE FOR THE ROLE OF THE NEURAL TUBE ON THE DEVELOPMENT OF SENSORY GANGLIA

As a corollary to the early segregation of autonomic and sensory cell lines, it was proposed (*12–14*) that, unlike autonomic progenitors, sensory neural precursors can survive only in close proximity to the neural tube. To verify this hypothesis, we performed experiments *in ovo* designed to examine more closely the influence of the axial structures on the development of neural crest into sensory ganglia in the chick embryo. This was done by devising an experimental system in which the somites were separated from the neural tube at the 28- to 30-somite stage at the level of somites 20 to 24 by a mechanical barrier (*9*). The barrier was a membrane of silastic inserted into a slit made over a length of about 4 somites (Fig. 6). The obstacle cut DRG anlagen into two unequal parts, a distal one, containing most of the neural crest cells, and a proximal one, in contact with the neural tube. Ten hours after insertion of the obstacle, the proximal crest cells (detected by immunostaining with HNK1) were still alive while the distal ones had totally disappeared (Fig. 6). Only necrotic remnants were seen on the external side of the silastic membrane. If the latter was impregnated with neural tube extract, survival—and, in some cases, differentiation into neurons—of a significant number of distally situated crest cells was observed for more than 30 hr after operation (Fig. 6). Liver extract incorporated into the membrane had no effect.

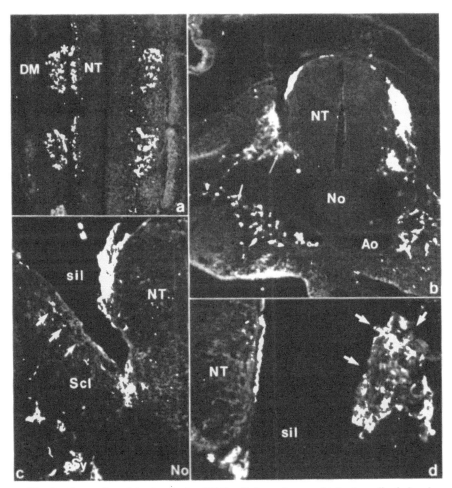

Fig. 6. Separation of the developing DRG cells from the neural tube by a silastic barrier. a, b: HNK-1 immunoreactivity in silastic membrane-implanted embryos 2 hr after operation. A: frontal section through somites 22–23 of a 32-somite embryo at the level of the DRG anlage. Note that most crest cells have been separated from the neural tube (NT) by the silastic membrane (asterisk) and remain on its somitic side. A small proportion of cells was left in contact with the tube. Neural crest cells on both sides of the membrane show HNK-1-positive immunoreactivity shortly after implantation. B: transverse section through somite 21, illustrating the same phenomenon as in A. The silastic membrane was put on the left side; the right side was unoperated. : A, 120; B, 92. C: cell death evidenced 10 hr after operation in the neural crest population separated from the CNS by a silastic membrane. Transverse section through somite 22 immunolabeled with the HNK-1 antibody. Operated side showing no HNK-1 positive cells on the distal side of the silastic membrane (sil) (arrows), while the cells left in contact with the NT as well

These results therefore suggested that specific factors produced by the CNS were responsible for the effect. One candidate for this function was the "brain-derived neurotrophic factor" (BDNF), recently extracted and purified from adult pig CNS by Barde *et al.* (*3*) on the basis of its action on survival and neurite outgrowth of chick DRG neurons in culture. Another was NGF, for which DRG have long been recognised to be a natural target. Moreover, NGF-like immunoreactivity has recently been demonstrated in the spinal cord and brain of certain mammalian foetuses (*1*), as has retrograde transport of ^{125}I-labelled NGF from the spinal cord of newborn rats to the DRG *via* the dorsal roots (*23*). However, incorporation of NGF into the silastic membrane barrier did not rescue distally located crest cells. In contrast, a membrane impregnated with BDNF and laminin allowed crest cell survival for more than 10 hr, although only in a small proportion of cases were crest cells found alive 30 hr after the operation (Kalcheim *et al.*, in preparation). Silastic membrane insertion never affected survival of somatic cells or their differentiation into cartilage and muscle. Likewise, sympathetic structures developed normally in the operated embryos. The developmental effects of the neural tube on sensory precursors is therefore clearly established and BDNF is probably one of the factors mediating it.

IV. CONCLUSIONS

In conclusion, the neural crest, a transitory structure which generates an impressive number of cell types and whose constitutive cells are endowed with migratory properties at early developmental stages, constitutes a fascinating developmental system (see *11*).

As far as PNS ontogeny is concerned, the results obtained so far —in the experiments described above and in others previously reported (see *11–14*)—support the contention that distinct cell lineages

---

←as future primary sympathetic ganglion cells (Sy) are positive. ×: 300. d: increased survival time of neural crest derived cells, on the somitic side of a silastic membrane impregnated with a neural tube extract. HNK-1 immunolabeling of a section from 4.5-day old embryos 30 hr after implantation. Many HNK-1 positive cells are evident on the somitic side of the membrane (arrows), in the region corresponding to the localisation of the DRG. ×: 320. From ref. *10* with permission.

become segregated early in crest cell ontogeny. Thier component cells, if not readily committed to express only a single phenotype, at least possess significant restrictions in their developmental potentials. Manipulating crest cell populations so that they were provided *in vivo* or *in vitro* with diverse microenvironments has led to the identification of at least three types of precursor cells (or three different lineages), including *sensory* (S), *sympathetic autonomic* (sA), and *enteric autonomic* (eA) precursors. These precursors are not equally distributed over the neuraxis and exhibit significantly different requirements for survival and differentiation. Whether they are derived from common ancestors and whether cells endowed with broader developmental potentials (*e.g.*, a cell whose progeny can belong to sensory and autonomic cell types) also exists in the migrating neural crest are still question marks. Moreover, the relationships between the precursors mentioned above, defined by the neuronal phenotypes they express, and the satellite, Schwann and glial cell family is a problem which requires further investigations. Those key questions will find an answer only if appropriate conditions are set up permitting the clonal development of neural crest cells.

The role of environmental factors in directing the expression of definite phenotypes in the neural crest cell population is of decisive importance. Elucidation of their nature and mode of action is, together with clonal analysis of crest cell ontogeny, the forthcoming priority in the endeavour to decipher the complex mechanisms leading from the neural crest to PNS ganglia.

SUMMARY

The neural crest, a transitory embryonic structure of the avian embryo, is at the origin of most of the ganglion cells of the PNS. Using the quail-chick chimaera system, a fate map of the PNS was constructed on the neural crest at the late neurula stage of the avian embryo. Changing the initial position of crest cells prior to the onset of migration did not significantly disturb PNS ontogeny, thus showing that the neural crest is endowed, at all levels of the neuraxis, with developmental potentialities broader than those expressed in normal development. Further *in vivo* and *in vitro* studies have shown that at least three types of neuronal cell lineages become segregated early in the neural crest; these correspond to the precursors of S, of sA and of eA. Each of them presents different survival and differentiation requirements. We have been able to demonstrate that the type S progenitors can survive and differentiate only if they are provided with growth factor(s) produced by the neural tube, this in agreement with the fact that all sensory ganglia whose neurons originate from the crest develop in close contact with the neural tube. Further studies are in progress to characterise each type of precursor cell and the factors on which they depend for their development.

Acknowledgments

The authors wish to acknowledge the assistance of E. Bourson, B. Henri, and A. Le Mouël for the preparation of this manuscript.

This work was supported by the Centre National de la Recherche Scientifique and by grants from the Institut National de la Santé et de la Recherche Médicale, the Ministère de la Recherche et de l'Industrie, the Fondation pour la Recherche Médicale Française and the Ligue Française contre la Cancer and by Basic Research Grant 1-866 from the March of Dimes Birth Defects Foundation.

REFERENCES

1 Ayer-Le Lièvre C. and Le Douarin, N.M. (1982). *Dev. Biol.* **94**, 291–310.
2 Ayer-Le Lièvre, C.S., Ebendal, T., Olson, L., and Sieger, A. (1983). *Med. Biol.* **61**, 296–304.
3 Barde, Y.A., Edgar, D., and Thoenen, H. (1982). *EMBO J.* **1**, 549–553.
4 Carr, V.McM. and Simpson, S.B. (1978). *J. Comp. Neurol.* **182**, 727–740.
5 D'Amico-Martel, A. and Noden, D.M. (1983). *Am. J. Anat.* **166**, 445–468.
6 Dupin, E. (1984). *Dev. Biol.* **105**, 288–299.
7 Fontaine-Pérus, J., Chanconie, M., and Le Douarin, N.M. (1987). *Dev. Biol.*, in press.
8 Iversen, L.L. (1973). *Br. Med. Bull.* **29**, 130–135.
9 Kalcheim, C. and Le Douarin, N.M. (1986). *Dev. Biol.* **116**, 451–466.
10 Le Douarin, N.M., Teillet, M.A., Ziller, C., and Smith, J. (1978). *Proc. Natl. Acad. Sci. U.S.A.* **75**, 2030–2034.
11 Le Douarin, N.M. (1982). *The Neural Crest.* Cambridge: Cambridge Univ. Press.
12 Le Douarin, N.M. (1984). In *Cellular and Molecular Biology of Neuronal Development*, ed. Black, I., pp. 3–28. New York: Plenum Press.
13 Le Douarin, N.M. (1986). *Science* **231**, 1515–1522.
14 Le Douarin, N.M. (1986). *Harvey Lecture Series*, pp. 137–186. Alan R. Liss Publ.: New York.
15 Le Lièvre, C.S., Schweizer, G.G., Ziller, C.M., and Le Douarin, N.M. (1980). *Dev. Biol.*

77, 362–378.

16 Newgreen, D.F., Jahnke, I., Allan, I.J., and Gibbins, I.L. (1980). *Cell Tissue Res.* **208**, 1–19.
17 Rohrer. H. (1985). *Dev. Biol.* **111**, 95–107.
18 Rothman, T.P., Gershon, M.D., and Holtzer, H. (1978). *Dev. Biol.* **65**, 322–341.
19 Schweizer, G., Ayer-Le Lièvre, C., and Le Douarin, N.M. (1983). *Cell Differ.* **13**, 191–200.
20 Xue, Z.G., Smith, J., and Le Douarin, N.M. (1985). *C. R. Acad. Sci. Paris* **300**, 483–488.
21 Xue, Z.G., Smith, J., and Le Douarin, N.M. (1985). *Proc. Natl. Acad. Sci. U.S.A.* **82**, 8800–8804.
22 Xue, Z.G., Smith, J., and Le Douarin, N.M. (1987). *Dev. Brain Res.*, in press.
23 Yip, H.K. and Johnson, Jr., E.M. (1984). *Proc. Natl. Acad. Sci. U.S.A.* **81**, 6245–6249.

7

EXPRESSION OF THE GENE CODING FOR LENS-SPECIFIC CRYSTALLINS IN THE NERVOUS TISSUES

T.S. OKADA

The National Institute for Basic Biology, Okazaki 444, Japan

Macromolecules predominantly present in a particular type of tissues serve as "markers" to follow the differentiation of a given cell type in terms of molecular changes. Crystallins are one representative of such marker molecules for the study of lens differentiation (*cf.* reviews: *4, 15*). Recent studies of the expression of δ-crystallin, a major protein of avian embryonic lenses, have shown that this particular protein is transcribed and translated at a low level in some non-lenticular tissues also. The expression is particularly consistent in embryonic nervous tissues including brains. Our observations of the transcription and translation of a gene coding for δ-crystallin are reviewed in this article. A search for biological implications of the presence of crystallins in brain is left for future studies.

I. A GENE CODING FOR δ-CRYSTALLIN IS TRANSCRIBED IN EMBRYONIC NERVOUS TISSUES

Although high expression of a gene coding for δ-crystallin (δ-gene) occurs only in lens in *in situ* developing chicken embryos, a low level transcription is found in a number of non-lenticular tissues such as neural retina, brains, limb buds, and others by assays utilizing cloned δ-crystallin cDNA as a probe (1, 5). It was claimed that δ-crystallin transcripts found in these non-lenticular tissues are incompletely processed and have higher molecular weights than mature mRNA found abundantly in lens cells. The level of transcription of δ-gene in brains was calculated as low as 10^{-5}–10^{-4} compared to in situ lens cells. The ectopic expression of δ-gene is detectable only in embryonic life and disappears in post-hatching stages.

II. EMBRYONIC NEURAL RETINA CELLS TRANSDIFFERENTIATE INTO LENS IN CELL CULTURE

It has been repeatedly demonstrated that cells of neural retina (NR) of both avian and mammalian embryos transdifferentiate very extensively into lens and pigment cells, when they are dissociated and maintained in the stationary culture in vitro (cf. reviews: 13, 14). Among several cell types which have already been differentiated in the NR tissue before culturing, glial cells have been known to transdifferentiate

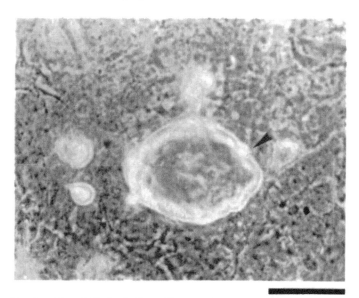

Fig. 1. A lentoid body (indicated by arrow) found in a 30-day culture of 8-day embryonic chicken NR cells. (bar: 50 μm).

into lens (8), although a possibility of early matrix cells as a progenitor of lens differentiation has not been completely excluded (7).

A visible lens differentiation from NR in culture is recognized by the formation of a number of lentoid bodies (Fig. 1). A dramatic rise in transcription and translation of δ-gene occurs in the process of lentoidogenesis *in vitro* ($2, 18$). Recently, we have assayed cellular changes associated with this process of transdifferentiation in respect to changes in the level of transcription activity of δ-gene. Since the promoter sequence of δ-gene responsible for the high level expression of this gene in lens cells is known (6), a hybrid gene was prepared, which is driven by δ-gene promoter and encodes bacterial chloramphenicol acetyl-transferase (CAT) enzyme. NR cells at various periods after the initiation of culturing were transfected with such hybrid genes and the expression was assayed by measuring CAT enzyme activity. The results demonstrate that cultured NR cells acquire an ability to support these exogenous genes with crystallin promoters well before the initial stage of transcription of endogenous crystallin genes. The stage of acquisition of such ability is assumed to be that of determination of lens differentiation (9).

III. TRANSCRIPTION AND TRANSLATION OF THE δ-CRYSTALLIN GENE ARE ACTIVATED IN CULTURED BRAIN CELLS

As will be described later, the presence of a minute amount of not only δ-crystallin mRNA (δ-mRNA), but also of δ-crystallin was demonstrated in chicken embryonic brains. When cells dissociated from the optic lobes and forebrains of 6-day-old embryos were cultured, δ-crystallin and δ-mRNA showed a significant increase from their *in situ* level. Immunohistological observation of cultures revealed that both flattened epithelial cells (glia-like cells) and cells with long dendritic or bipolar processes are δ-producers (16). No lentoidogenesis occurred. Therefore, in this case, the synthesis of "lens-specific" δ-crystallin was stimulated without any accompanying morphological characteristics such as authentic lens differentiation. It was suggested that δ-crystallin can be synthesized in some cells with glial or neuronal phenotypes.

IV. TRANSITORY EXPRESSION OF THE δ-CRYSTALLIN GENE IN EMBRYONIC BRAINS AND ASSOCIATED STRUCTURES OCCURS IN *IN SITU* EMBRYOS

Rathke's pouch, a rudimentary structure of adenohypophysis, is known as a unique place with a relatively high content of δ-crystallin (*3*). This is only one non-lenticular structure shown to be positive to this protein by immunofluorescence technique (Fig. 2). The expression of δ-crystallin in Rathke's pouch is in the range of 0.1–1% of the level in the lens cells (*17*).

When high-sensitive immunohistological staining by the PAP method was applied to sections of chicken embryos, positive staining to anti-δ-crystallin also occurred in brains and spinal cords. The presence of δ-crystallin in brains, however, was transitory, and could be detected only in embryos from the early half of the embryonic period. The localization was also very limited to some cells in slightly ventrally located sectors of the tissues (Fig. 3). The observations were reproducible, so that we can conclude that the δ-crystallin gene is, though at a

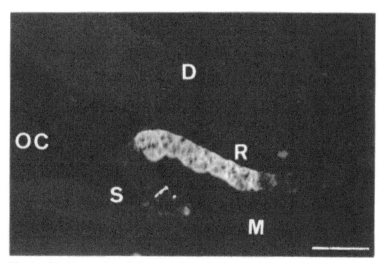

Fig. 2. Median section of the adenohypophysis of a 3.5-day-old chicken embryo stained with antibodies against δ-crystallin using indirect immunofluorescence. D, diencephalon; M, mesenchymal tissue; OC, oral cavity; R, Rathke's pouch; S, Sessel's pouch. (bar; 100 μm). (Ueda and Okada, 1986)

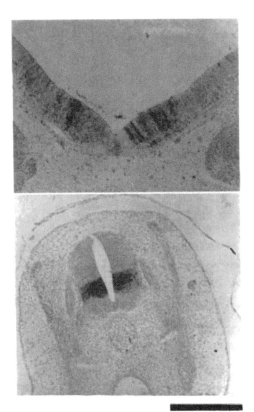

Fig. 3. The localization of δ-crystallin in the hind brain (upper) and in the spinal cord (lower) of a 3-day chicken embryo, as revealed by the immunohistological technique with PAP. (bar; 200 μm).

low level, transcribed and translated in a particular part of developing embryonic nervous systems.

V. EXPRESSION OF THE δ-CRYSTALLIN GENE IN BRAINS OF TRANSGENIC MICE

The expression of δ-gene was analyzed in various tissues of transgenic mice which were grown after the injection of this particular gene into egg male pronucleus. In both of two such mice so far studied, the expression of δ-gene occurred exclusively in the lens and in the brain. In the latter, the cells immunohistologically positive in histological

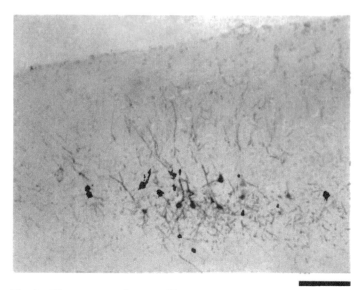

Fig. 4. The presence of δ-crystallin in the pyramidal neurons of piriform cortex of a transgenic mouse incorporated with δ-gene, as revealed by immunohistological technique with PAP. (bar; 100 μm). (Kondoh *et al.*, 1987)

sections were the pyramidal neurons of the piriform cortex (*11*) (Fig. 4).

VI. EXPRESSION OF THE δ-CRYSTALLIN GENE IN CHIMERIC EMBRYOS

We have established a number of transformed clonal lines of teratocarcinoma cells (*8*) and of embryonic stem cell lines (EK cells) with the intention of studying the regulatory mechanisms of the expression of δ-gene in the process of differentiation of these "multipotent" cells. As long as the cells grew *in vitro* in undifferentiated state, no expression of the exogenous δ-genes occurred. In the solid tumors (teratomas) produced by injecting the transformed cells *in vitro* into syngenic mice, the expression of δ-gene was found in various differentiated tissues such as muscles, endothelial cells, columnar epithelium cells and others, depending on each clonal line. Such δ-crystallin expression in the "ectopic" tissues was explained by tissue-specific variation of the chromosonal activities at the different integration sites (*10*).

A number of chimeric embryos were produced by introducing the transformed EK cells into blastocysts of different strains. In these cases, the expression occurred in lens and brains. It should be remarked that most of the positive cells were found in the ventral half of the brain, which coincided with the δ-crystallin containing area of the embryonic chicken brains *in situ*. Therefore, regulation systems of the mouse handle the heterologous δ-crystallin gene with a high stringency to mimic the chicken system. When we speak of "high stringency," the ectopic expression of "lens-specific" δ-crystallin in brains is also included.

SUMMARY

δ-crystallin, one of the so-called "lens-specific" proteins, is found in chicken embryonic nervous tissues, although at very low level compared to lens cells. The observation was substantiated also by the expression of δ-gene introduced into mice embryos by two different gene transfer techniques. The transcription and translation of δ-gene in nervous tissues are much stimulated when the cells are dissociated and transferred into cell culture conditions. Only in neural retina cells is such stimulation accompanied by authentic lens differentiation. The possible function of δ-crystallin in nervous tissues is not known.

Acknowledgment

The work reviewed in this article was conducted in the Institute for Biophysics, University of Kyoto with the author's previous collaborators. Among them, the author is particularly indebted to Dr. Hisato Kondoh for advice in preparing the manuscript. He also thanks Ms. Toyoko Tsuge for secretarial help.

REFERENCES

1 Agata, K., Yasuda, K., and Okada, T.S. (1983). *Dev. Biol.* **100**, 222–226.
2 Araki, M. and Okada, T.S. (1977). *Dev. Biol.* **60**, 278–286.
3 Barabanov, V.M. (1977). *Dokl. Acad. Nauk.* **234**, 195–198.
4 Clayton, R.M. (1973). In *The Eye*, eds. Davison, H. and Graham, L.T., vol. 5, pp. 399–494. New York, London: Academic Press.
5 Clayton, R.M., Thomson, I., and dePomerai, D.I. (1979). *Nature* **282**, 268–269.
6 Hayashi, S., Kondoh, H., Yasuda, K., Soma, G., Ikawa, Y., and Okada, T.S. (1955). *EMBO J.* **4**, 2201–2207.

7 Kondoh, H., Takagi, S., Nomura, K., and Okada, T.S. (1983). *Roux' Arch. Dev. Biol.* **192**, 256–261.

8 Kondoh, H., Takahashi, Y., and Okada, T.S. (1984). *EMBO J.* **3**, 2009–2014.

9 Kondoh, H., Ueda, Y., Hayashi, S., Okazaki, K., Yasuda, K., and Okada, T.S. (1987). *Cell Differ.* **20**, 203–207.

10 Kondoh, H., Hayashi, S., Takahashi, Y., and Okada, T.S. (1986). *Cell Differ.* **19**, 151–160.

11 Kondoh, H., Katoh, K., Takahashi, Y., Fujisawa, H., Yokoyama, M., Kimura, S., Katsuki, M., Saito, M., Nomura, T., Hiramoto, Y., and Okada, T.S. (1987) *Dev. Biol.* **120**, 177–185.

12 Moscona, A.A. (1986). *Curr. Top. Dev. Biol.* **20**, 1–19.

13 Okada, T.S. (1980). *Curr. Top. Dev. Biol.* **16**, 349–380.

14 Okada, T.S. (1983). *Cell Differ.* **13**, 177–183.

15 Piatigorsky, J. (1984). *Cell* **38**, 620–621.

16 Takagi, S. (1986). *Roux' Arch Dev. Biol.* **195**, 15–21.

17 Ueda, Y. and Okada, T.S. (1986). *Cell Differ.* **19**, 179–187.

18 Yasuda, K., Okuyama, K., and Okada, T.S. (1983). *Cell Differ.* **12**, 85–92.

NEURAL PLASTICITY

8

SYNAPTIC PLASTICITY IN THE CEREBELLAR CORTEX AS A MEMORY PROCESS FOR MOTOR LEARNING

MASAO ITO

Department of Physiology, Faculty of Medicine, University of Tokyo, Tokyo 113, Japan

During the past two decades, four types of synaptic plasticity have been demonstrated to exist in the central nervous system. The long-term potentiation (LTP) prevails in the hippocampus (2), but it appears also to occur in the neocerebral cortex. LTP is characterized by its associative nature that signal transmission at a synapse of a hippocampal neuron is facilitated when its activity is associated with effective transmission in other synapses on the same neuron. The LTP appears to be a cellular basis of cognitive memory and learning performed in the cerebral cortex. The second type is the long-term depression (LTD) found in the cerebellar cortex (8). This takes place when a synapse supplied by a parallel fiber axon of a granule cell to a Purkinje cell is activated in conjunction with another powerful synapse from a climbing fiber. LTD is also associative, but the direction of the synaptic efficacy change in LTD is opposite to that in LTP. The LTD is presumed to be the basis of motor learning performed by the cerebellum. The third is the sensitization found in *Aplysia ganglia*, in which transmission at a synapse is facilitated by a brief action of another synapse straddling the former (11). The sensitization is essentially of a non-associative type, but its associative operation is also known. It is instrumental in sensitization of gill-withdrawal reflex in *Aplysia*. The fourth type is synaptic sprouting observed in the spinal cord, brainstem and hippocampus. It occurs prominently at terminals of cerebral coticofugal fibers in the red nucleus after lesion of cerebellar inputs to the red nucleus (*16*). Even a remote cause such as cross-union of peripheral nerves evokes synaptic sprouting in the red nucleus, which may be a cellular basis of compensation for peripheral motor dysfunction.

This chapter reviews the current knowledge of the LTD.

I. INDUCTION OF LTD

LTD was first demonstrated to occur with indirect stimulation of parallel fibers *via* mossy fiber afferents in conjunction with stimulation of climbing fibers, and recording from individual Purkinje cells with an extracellular microelectrode (*8*). The responsiveness of floccular Purkinje cells was tested by applying single pulse stimuli to a vestibular nerve that projects to the cerebellar flocculus as mossy fiber afferents. These stimuli excited Purkinje cells with a latency of 3–6 msec, presumably *via* granule cells and parallel fibers. Conjunctive stimulation of climbing fibers and mossy fibers was performed by applying 4-Hz stimuli to the inferior olive and 20-Hz stimuli to a vestibular nerve simultaneously for a period of 25 sec. Thus, 100 climbing fiber impulses and 500 vestibular mossy fiber impulses were given to the flocculus. A stimulation of 4 Hz was chosen, as it is about the maximum climbing fiber discharge observed in alert rabbits; 20 Hz is a moderate frequency for discharge evoked in vestibular nerve fibers by head rotation. With these parameters, conjunctive stimulation effectively depressed the vestibular mossy fiber responsiveness of floccular Purkinje cells. Although the depression diminished in about 10 min, there was usually a slow phase of weak depression that lasted for more than 1 hr. This effect was specific to the vestibular nerve involved in the conjunctive stimulation; no depression occurred in the responsiveness to inputs from the opposite vestibular nerve. Therefore, the effect of a conjunctive stimulation is not due to general depression in Purkinje cells, but is specific to a site or sites in the mossy fiber-granule cell-Purkinje cell pathway involved in the conjunctive stimulation. Conjunctive stimulation of climbing fibers and mossy fibers did not affect the responsiveness of granule cells or Golgi cells to mossy fiber inputs, as indicated by field potentials recorded in the granular layer and white matter of the flocculus. It was thus concluded that the site of LTD was at parallel fiber-Purkinje cell synapses.

LTD was then shown to be induced by direct stimulation of parallel fibers in conjunction with climbing fibers, and by recording of mass field potentials from the molecular layer of the cerebellar cortex (7). Parallel fibers were stimulated directly through a glass microelectrode placed in the molecular layer. Application of brief pulses (0.2 msec duration) of 20–60 μA excited a narrow beam of parallel fibers along which double-peaked negative field potentials were recorded with another microelectrode. The first peak represents volleys conducted along parallel fibers, and the second peak represents the postsynaptic excitation thereby evoked in dendrites of Purkinje and other cortical cells. When the parallel fiber stimuli were paired with stimuli to the inferior olive with a time delay of 2–6 msec, and repeated at a rate of 4 pulses/sec for 1 or 2 min, the second peak often exhibited a small but significant depression that lasted for more than 1 hr. The depression however, was usually small (10–20%) and was often difficult to detect, probably because of the heterogeneous origin of the mass field potentials from Purkinje, basket, stellate, and Golgi cells.

When the recording microelectrode is placed in the Purkinje cell layer, activation of single Purkinje cells by parallel fiber volleys can be observed (5). Conjunctive stimulation of a parallel fiber beam and the inferior olive at 1–4 Hz regularly produced a prominent depression in the responsiveness of Purkinje cells to parallel fiber stimulation. The depression lasted for a period of more than 1 hr. Some Purkinje cells were excited through two separate parallel fiber beams. When one of the two beams was stimulated in conjunction with the inferior olive, the responsiveness of the Purkinje cell to that parallel fiber beam was specifically depressed; the responsiveness to the other beam remained unchanged.

Intradendritic recording from a Purkinje cell in a cerebellar slice preparation also revealed the LTD. An excitatory synaptic potential (EPSP) evoked by stimulation of parallel fibers was reduced appreciably (by 20–50%) in peak amplitude or the initial rising slopes after conjunctive activation with climbing fibers at 4 Hz for 25 sec. Stimulation of climbing fibers alone had no such effect. Stimulation of parallel fibers alone even induced slight potentiation of the parallel fiber-induced EPSP. Thus, conjunctive stimulation of both climbing fibers and parallel fibers is requisite for induction of LTD.

The timing of parallel fiber and climbing fiber stimulations to induce LTD has been found to have a relatively wide latitude. Parallel fiber stimulations falling during the period between 20 msec prior to and 150 msec subsequent to the stimulation of climbing fibers are nearly equally potent in inducing LTD (Ekerot and Kano, personal communication). Even those occurring at 250 msec after climbing fiber stimulation induced LTD, though with less probability.

II. INVOLVEMENT OF Ca^{2+} IN LTD

Involvement of Ca^{2+} in LTD has been suggested by the finding that LTD was abolished when climbing fiber inputs were conditioned with postsynaptic inhibition of Purkinje cell dendrites through stellate cells (5). Since stellate cell inhibition depresses both the Ca^{2+}-spikes and subsequent Ca^{2+}-dependent plateau potentials induced by climbing fiber impulses, the above observation suggests that Ca^{2+} influx plays an essential role in inducing LTD.

More direct evidence for the role of Ca^{2+} influx has been obtained by intradendritic injection of Ca^{2+} chelator, EGTA (Sakurai, personal communication). Iontophoretic injection of EGTA through an electrode containing EGTA plus K-acetate abolished the LTD, whereas control injection of acetate ions did not affect it.

III. INVOLVEMENT OF GLUTAMATE RECEPTORS IN LTD

Ito *et al.* (8) replaced parallel fiber stimulation with iontophoretic application of glutamate, which is a putative neurotransmitter of parallel fibers. The application of glutamate and aspartate through a multi-barrelled pipette normally excited Purkinje cells. The application of glutamate in conjunction with 4 Hz stimulation of the inferior olive effectively depressed the glutamate sensitivity of Purkinje cells; aspartate sensitivity, tested as control, was depressed to a much lesser degree.

The depression diminished in about 10 min and was followed by a slow depression lasting for 1 hr. The depression of glutamate sensitivity was obtained only when there was an initially high glutamate sensitivity, higher than control aspartate sensitivity, suggesting that the electrode was close to parallel fiber-Purkinje cell junctions. Another important finding (9) was that when glutamate was applied to a dendritic region of a Purkinje cell in conjunction with the stimulation of climbing fibers, the responsiveness of the Purkinje cell to stimulation of a parallel fiber bundle that passed the dendritic site of glutamate application underwent LTD, just like LTD obtained with conjunctive stimulation of parallel fibers and climbing fibers. Similar depression of parallel fiber-Purkinje cell transmission was induced by application of quisqualate, a glutamate agonist, in conjunction with climbing fibers, but of neither aspartate nor kinate, even though they also activated Purkinje cells effectively. Since another glutamate agonist, N-methyl-D-aspartate (NMDA) does not activate Purkinje cells effectively, its involvement in LTD is doubtful. Therefore, it is concluded quisqualate-specific glutamate receptors at parallel fiber-Purkinje cell synapses are responsible for LTD.

Effect of conjunctive application of glutamate with climbing fiber impulses in inducing LTD was abolished when the excitatory effect of glutamate on a Purkinje cell was blocked by iontophoretic application of kynurenic acid (10). These observations indicate that the LTD is due to desensitization of glutamate receptors that takes place under action of neurotransmitter of parallel fibers.

IV. LINKAGE FROM Ca^{2+} INFLUX TO GLUTAMATE DESENSITIZATION

Desensitization of acetylcholine receptors is facilitated by Ca^{2+}. However, there is no available evidence suggesting that desensitization of glutamate receptors is likewise facilitated by Ca^{2+}.

Desensitization of glutamate receptors in cerebellar synaptosomes is facilitated by a relatively high concentration of cGMP (15). cGMP and cGMP-dependent protein kinase are contained specifically in Purkinje cells (12), and the level of cGMP in Purkinje cells is elevated after activation of climbing fibers (1). It is also known that Ca^{2+} activates guanylate cyclase in cerebellar slices (13). These data suggest that climbing fiber impulses enhance cGMP content in Purkinje cell dendrites, thereby facilitating desensitization of glutamate receptors of the parallel-fiber neurotransmitter. However, further pharmacological investigation is needed to test this possibility.

COMMENTS

LTD is now established as a unique and characteristic synaptic plasticity in the cerebellum. The possibility that LTD plays a key role in cerebellar functions has been substantiated by studies of the vestibulo-ocular reflex (VOR). The hypothesis that the flocculus acts as a center of adaptive modification of the VOR utilizing LTD as a memory process has been supported by lesion studies, recording of Purkinje cell signals, and computer simulation (6). The basic concept thus emerges that the cerebellum constitutes an adaptive control machine that assists motor and autonomic systems operating in other parts of the central nervous system.

SUMMARY

The LTD is a special type of synaptic plasticity present in the cerebellar cortex. It occurs at a synapse from a parallel fiber to a Purkinje cell when activated in conjunction with a climbing fiber converging onto the same Purkinje cell. Evidence so far suggests that the LTD is triggered by Ca^{2+} inflow into Purkinje cell dendrites accompanying climbing fiber impulses and eventually is brought about by desensitization of quisqualate-specific glutamate receptor molecules in Purkinje dendritic spines. Whether Ca^{2+} ions act directly on glutamate receptors or indirectly *via* some second messenger process such as activation of guanylate cyclase remains yet to be determined.

REFERENCES

1 Biggio, G. and Guidotti, A. (1976). *Brain Res.* **107**, 365–373.
2 Bliss, T.V.P. and Lomo, T. (1977). *J. Physiol.* **232**, 331–356.
3 Brindley, G.S. (1964). *IBRO Bull.* **3**, 80.
4 Changeux, J.-P., Klarsfeld, A., and Heidman, T. (1987). In *The Neural and Molecular Bases of Learning*, ed. Changeux, J.-P. and Konishi, M., pp. 31–84. New York: John

Wiley & Sons.

5 Ekerot, C.-F. and Kano, M. (1985). *Brain Res.* **342**, 357–360.

6 Ito, M. (1984). *The Cerebellum and Neural Control.* New York: Raven Press.

7 Ito, M. and Kano, M. (1982). *Neurosci. Lett.* **33**, 253–258.

8 Ito, M., Sakurai, M., and Tongroach, P. (1982). *J. Physiol.* **324**, 113–134.

9 Kano, M. and Kato, M. (1987). *Nature* **325**, 276–279.

10 Kano, M. and Kato, M. (1987). *J. Physiol.*, in press.

11 Klein, M. and Kandel, E.R. (1980). *Proc. Natl. Acad. Sci. U.S.A.* **77**, 6912–6916.

12 Lohmann, S.M., Walter, U., Miller, P.E., Greengard, P., and Camilli, P.D. (1981). *Proc. Natl. Acad. Sci. U.S.A.* **78**, 653–657.

13 Ohga, Y. and Daly, J.W. (1977). *Biochim. Biophys. Acta* **498**, 61–75.

14 Sakurai, M. (1987). *J. Physiol.*, in press.

15 Sharif, N.A. and Roberts, P.J. (1980). *Eur. J. Pharmacol.* **61**, 213–214.

16 Tsukahara, N. (1981). *Annu. Rev. Neurosci.* **4**, 351–379.

9

QUANTAL ANALYSIS OF POTENTIATION OF SYNAPTIC TRANSMISSION IN THE HIPPOCAMPUS

CHOSABURO YAMAMOTO, MASATO HIGASHIMA, AND
SATSUKI SAWADA

Department of Physiology, Faculty of Medicine, Kanazawa University, Kanazawa 920, Japan

After a brief train of tetanic stimulation is delivered to an excitatory pathway in the hippocampus, synaptic transmission through the tetanized pathway is facilitated for a long period (*4, 7, 10*). This phenomenon is called long-term potentiation (LTP). Since the magnitude of LTP decreases concomitantly with reduction in the ability of learning in old animals (*1*) and under the action of a blocker of receptors in the brain (*16*), LTP has been considered to have relation with memory. To study the synaptic mechanism of the LTP seems important to understand plasticity in the brain.

The question remains unanswered whether the changes underlying LTP take place in the presynaptic terminals or postsynaptic neurons. Bliss and his collaborators observed increases in L-glutamate (Glu) and L-aspartate release from the hippocampus with development of LTP (*3*). Sastry and his associates reported data suggesting that hyperpolarization of presynaptic terminals occurred with LTP (*17*). These authors, therefore, consider the presynaptic terminals as the site of primary changes underlying LTP. Recently, Malenka *et al.* (*12*) showed that excitatory postsynaptic potentials (EPSPs) in subfield CA1 increases in the presence of phorbol ester. This effect of phorbol ester is sustained even if the ester is removed from the perfusing solution. In addition, they found that once a tissue is exposed to phorbol ester, tetanic stimulation no longer induces LTP in the tissue. From these observations, they consider activation of protein kinase C triggers LTP. Moreover, they assert the presynaptic origin of LTP because phorbol ester does not increase responses of CA1 neurons to Glu, although this seems to be unconvincing evidence.

On the other hand, Lynch and his associates reported that injection of EGTA into the postsynaptic neurons blocks LTP (11). Malinow and Miller (13) reported that hyperpolarization of postsynaptic neurons blocks the generation of LTP. These findings suggest the postsynaptic origin of LTP. Moreover, LTP has a strange property called cooperativity (15). When an input to a group of hippocampal neurons is tetanized at low intensities, no LTP takes place. But, simultaneous tetanization of another input at high intensity induces LTP in the former input. This means that LTP occurs only when input volley is accompanied by an intense excitation in postsynaptic neurons. This also suggests the postsynaptic origin of LTP.

I. APPLICATION OF QUANTAL ANALYSIS TO HIPPOCAMPUS

We are now attempting to determine which is the cause of LTP, an increase in the amount of neurotransmitter released from the presynaptic terminals or an increase in the responsiveness of the postsynaptic neurons to the transmitter, because this issue is essential to clarify the mechanism of LTP. For this purpose, we need to apply the quantal analysis technique. The principle of the quantal analysis technique is very simple (9). We stimulate a constant number of input fibers regularly many times and record all of the evoked EPSPs. From the fluctuation of the EPSP amplitudes, we can estimate quantal amplitude (q), namely, the amplitude of EPSPs induced by a single quantum of transmitter. We can also calculate the mean number of quanta (m) liberated by single impulses.

Although simple in principle, the quantal analysis technique is difficult to appliy to hippocampal synapses. The difficulty is 2-fold. First, when we repeat stimulation, we must be sure that we are stimulating a constant number of input fibers every time. This is because the fluctuation of the EPSP amplitude must not be due to the fluctuation of the number of excited fibers. When the authors used electric stimulation, however, they could not be confident that they were stimulating a constant number of input fibers. The second difficulty of applying the quantal analysis technique is the presence of inhibitory postsynaptic potentials (IPSPs). In the hippocampus, EPSPs evoked by electric stimulation of input fibers are usually accompanied by IPSPs (*18*). The IPSPs make accurate measurement of EPSP amplitudes almostimpossible.

In order to overcome these difficulties, we devised a method, the principle of which is shown in Fig. 1 (*6, 19*). A microelectrode is inserted into a CA3 neuron in the hippocampus to record intracellular potentials. Then, a micropipette filled with Glu solution is inserted repeatedly into the granule cell layer. When the tip of the Glu pipette encounters a granule cell whose axon makes synaptic contact with the impaled neuron, ejection of Glu from the pipette is expected to activate the granule cell and thereby to induce EPSP trains in the particular CA3 neuron. The position of the tip of the pipette is adjusted meticulously so that Glu pulses induce EPSP trains at the lowest intensity. Since the spread of Glu ejected with short pulses at low intensities seems to be very limited (*20*), we may expect to activate only one granule cell to make a synapse with the impaled neuron.

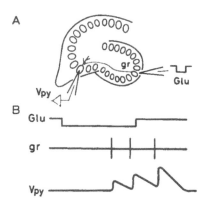

Fig. 1. Induction of an EPSP train by a Glu pulse to the dentate gyrus. Details in text.

II. PROPERTIES OF EPSPs INDUCED BY GLU PULSES TO GRANULE CELLS

Figure 2 shows EPSP trains actually recorded from a CA3 neuron in a thin section of the guinea pig hippocampus in response to Glu pulses to the dentate gyrus. At the most effective sites, Glu pulses induced trains composed of 3 EPSPs (records B1 and B5). These EPSPs exhibited marked frequency potentiation: the second EPSP was larger than the first one and the third EPSP was larger than the second one. With removal of the tip of the Glu pipette from the most effective site at 5 μm steps, the latency of the EPSP trains was prolonged and the number of EPSPs decreased successively in an all-or-none manner. After 15 μm of removal, EPSPs were no longer elicited. No smaller EPSPs or IPSPs were revealed after disappearance of the large EPSPs. This means that the extent of spread of Glu is limited within 20 μm, and that among granule cells making synaptic contact with the impaled neuron only one cell was activated by the Glu pulses. Figure 2 also indicates that the EPSPs were not contaminated by IPSPs.

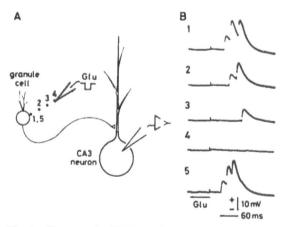

Fig. 2. Decreases in EPSP number with removal of the tip of a Glu pipette by 5 μm steps. A shows a diagram of experimental design. B1 shows an EPSP train recorded when the tip of the Glu pipette was positioned at the most sensitive spot. B2, B3, and B4 show potentials recorded after removal of the Glu pipette 5, 10, and 15 μm from the most sensitive spot, respectively. B5 shows an EPSP train after the Glu pipette was brought back to the most sensitive spot.

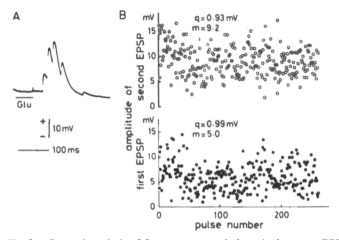

Fig. 3. Quantal analysis of frequency potentiation. A shows an EPSP train recorded from a CA3 neuron. In B, the amplitudes of the first (lower graph) and the second EPSPs (upper graph) in each train recorded from the same neuron are plotted successively. A solution containing Ca^{2+} at 2.4 mM and Mg^{2+} at 1.3 mM was used. In this and in the following illustrations, the amplitudes of EPSPs were corrected for non-linear summation on the assumption that the reversal potential for the EPSP was 0 mV (14).

TABLE I

Summary of Results of Quantal Analysis of EPSPs Elicited in Standard Solution and High Calcium Solution (6)

	Mean unitary EPSP (mV)	m	q(mV)	g_q ($\times 10^{-10}$S)
Standard solution	2.5±1.0	4.4±1.9	0.61±0.17	6.4±2.0
High calcium solution	7.8±3.5*	16.0±9.1*	0.57±0.26	9.3±4.5

These values were calculated for the first EPSPs in individual trains. Number of neurons was 11 in each solution. All values given as mean±S.D. *$p<0.05$, as compared with the values in a standard solution.

In the experiment in Fig. 3, we examined the effects of frequency potentiation on the values of q and m. Because of frequency potentiation, the amplitude of the second EPSP was larger than that of the first in most trains. The amplitude of the first and second EPSPs was measured and plotted in B, and the mean number of released quanta were estimated at 5.0 for the first EPSP and 9.2 for the second EPSP. On the other hand, the quantal amplitudes were 0.99 mV and 0.93 mV for the first and second EPSPs, respectively. This indicates that the potentiation of the second EPSP is explainable solely by an increase in the number of quanta liberated by impulses.

In a solution containing calcium ions at 4.8 mM, the mean amplitude of the first EPSPs was 7.8 mV (Table I). This was about 3 times as large as that in the standard solution containing calcium ions at 2.4 mM. The mean number of released quanta was 3.5 times as large in the high calcium solution. On the other hand, there was no significant difference in the mean value of the quantal amplitude. Therefore, the increase in the EPSP amplitude may be explained by increases in the number of released quanta rather than by increases in sensitivity of postsynaptic receptors to the transmitter. Frequency potentiation and high calcium concentration are known in neuromuscular junctions and in other peripheral synapses to potentiate EPSPs by increasing the mean number of quanta liberated by impulses without changing quantal amplitude (2, 5, 10). Our results are in full agreement with the findings in the peripheral synapses. Our method of quantal analysis seems, therefore, to be reliable for application to the study of LTP.

III. QUANTAL ANALYSIS OF LTP INDUCED BY PHORBOL ESTER

According to Malenka *et al.* (12), we took the potentiation of EPSPs by phorbol ester as a model of LTP. At the outset, we examined the effect of phorbol ester on the field potentials evoked in CA3 by electric stimulation of the granule cells. As reported by Malenka *et al.* (12), the amplitudes of EPSPs gradually increased during administration of phorbol diacetate. It took about 20 min until the amplitude became stabilized. In the following analysis, therefore, the tissue was perfused with a solution containing phorbol diacetate for more than 20 min before the data were collected.

Figure 4 shows the data from an experiment. Before and during administration of 0.1 μM phorbol diacetate, trains, each composed of 2 EPSPs, were elicited at 0.5 Hz (except for a short period just preceding phorbol diacetate). This illustration, in which the amplitudes of the first EPSPs are plotted against time, shows that the EPSP amplitude fluctuated from train to train. On the average, the EPSP was augmented during administration of phorbol diacetate. From the mean

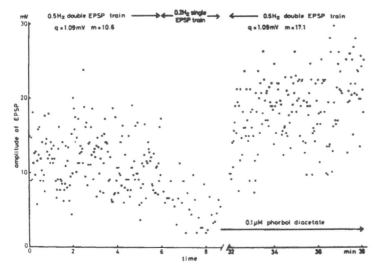

Fig. 4. Fluctuation of the amplitudes of EPSPs before and during administration of phorbol diacetate. Trains composed of 2 EPSPs were induced at 0.5 Hz in the control period and during administration of 0.1 μM phorbol diacetate except for a short period just before phorbol diacetate administration during which single EPSPs were induced at 0.2 Hz. The amplitudes of the first EPSPs in successive trains were measured and plotted against time. A solution containing 3.6 mM Ca^{2+} and 1.0 mM $MgSO_4$ was used.

and variance of the EPSP amplitude, the amplitude of quantal EPSP and number of released quanta were estimated. In control solution, the value of q was 1.09 mV. During phorbol diacetate administration, this remained unaltered. On the other hand, the value of m increased from 10.6 to 17.1, indicating that the potentiation induced by phorbol ester resulted from an increase in the release of transmitter.

In calculating the values of m and q, it was assumed that the amplitudes of EPSPs fluctuated according to Poisson's law. This is examined for the neuron in Fig. 4. In Fig. 5, dots show the theoretically expected distribution of the amplitudes of EPSPs calculated with the values of m and q in Fig. 4 on the assumption that Poisson's law is applicable. The entire distribution of amplitudes of observed EPSPs is shown by rectangles. It is evident that the actual distribution of the EPSP amplitudes largely agreed with the theoretically expected one.

We have successfully obtained data from 7 neurons. All of them showed significant potentiation of EPSPs under the action of phorbol

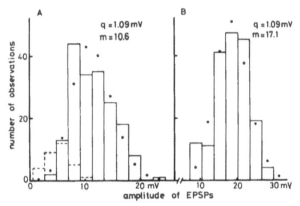

Fig. 5.   Distribution of amplitudes of EPSPs in the preceding illustration. A: control. B: phorbol diacetate. Dots, theoretically expected distribution of EPSP amplitudes; rectangles, distribution of amplitudes of observed EPSPs.

diacetate at 0.1–0.45 μM. In 5 of them, potentiation of EPSPs was accompanied by increases in the value of m with no considerable changes in the value of q. In the remaining 2 neurons, the value of q decreased with marked increases in the value of m. In no neurons did the value of q increase, indicating that an increase in sensitivity of postsynaptic receptors to neurotransmitter cannot explain the facilitation of transmission under the action of phorbol ester.

IV.   CONTROL OF TRANSMITTER RELEASE BY POSTSYNAPTIC ELEMENTS?

Experiments are now in progress in our laboratory to see whether the LTP induced by tetanic stimulation is also of presynaptic origin. In one pilot experiment, we obtained affirmative results. If the data are confirmed in more neurons, it will be asked how the cooperativity can be explained. The possibility may be suggested that postsynaptic neurons can control liberation of the neurotransmitter from presynaptic boutons. This will be one of the central questions to be studied in the next step.

SUMMARY

A method was developed to study EPSPs recorded from CA3 neurons in transverse sections of the guinea pig hippocampus with quantal analysis technique. For this purpose, a single granule cell was activated by brief L-Glu pulses and EPSPs were recorded from a CA3 neuron which was innervated by the activated granule cell. From the mean amplitudes of the EPSPs and the magnitude of fluctuation of the EPSP amplitudes, the quantal size (q) and the mean number (m) of quanta released by single impulses were calculated. Increases in EPSP amplitude by preceding impulses or in a high calcium solution were accompanied by increases in values of m without any significant changes in the value of q. Potentiation induced by phorbol diacetate was also accompanied by increases in the values of m. It was concluded that phorbol ester causes an increase in release of transmitter and thereby induces LTP.

Acknowledgment

This study was supported by a grant from the Ministry of Education, Science and Culture of Japan (61131003).

REFERENCES

1 Barnes, C.A. (1979). *J. Comp. Physiol. Psychol.* **93**, 74–104.
2 Blackman, J.G., Ginsborg, B.L., and Ray, C. (1963). *J. Physiol.* **167**, 402–415.
3 Bliss, T.V.P., Douglas, R.M., Errington, M.L., and Lynch, M.A. (1986). *J. Physiol.* **377**, 391–408.
4 Bliss, T.V.P. and Gardner-Medwin, A.R. (1973). *J. Physiol.* **232**, 357–374.
5 Dodge, F.A. and Rahamimoff, R. (1967). *J. Physiol.* **193**, 419–432.
6 Higashima, M., Sawada, S., and Yamamoto, C. (1986) *Neurosci. Lett.* **68**, 221–226.
7 Higashima, M. and Yamamoto, C. (1985). *Exp. Neurol.* **90**, 529–539.
8 Higashima, M. and Yamamoto, C. (1986). *Neurosci. Res.* **3**, 660–665.
9 Hubbard, J.I., Llinas, R., and Quastel, D.M.J. (1969). *Electrophysiological Analysis of Synaptic Transmission.* London: Arnold.
10 Katz, B. and Miledi, R. (1968). *J. Physiol.* **195**, 481–492.
11 Lynch, G., Larson, J., Kelso, S., Barrionuevo, G., and Schottler, F. (1983). *Nature* **305**, 719–721.
12 Malenka, R.C., Madison, D.V., and Nicoll, R. (1986). *Nature* **321**, 175–177.
13 Malinow, R. and Miller, J.P. (1986). *Nature* **320**, 529–530.
14 Martin, A.R. (1955). *J. Physiol.* **130**, 114–122.
15 McNaughton, B.L., Douglas, R.M., and Goddard, G.V. (1978). *Brain Res.* **157**, 227–293.
16 Morris, R.G.M., Anderson, E., Lynch, G.S., and Baudry, M. (1986). *Nature* **319**, 774–776.
17 Sastry, B.R. (1982). *Life Sci.* **30**, 2003–2008.
18 Yamamoto, C. (1972). *Exp. Brain Res.* **14**, 423–435.
19 Yamamoto, C. (1982). *Exp. Brain Res.* **46**, 170–176.
20 Yamamoto, C. and Sawada, S. (1982). *Brain Res.* **235**, 358–362.

10

PROTEIN PHOSPHORYLATION, A BIOCHEMICAL APPROACH TO NEURONAL PLASTICITY

USHIO KIKKAWA, KOUJI OGITA, AND
YASUTOMI NISHIZUKA

Department of Biochemistry, Kobe University School of Medicine, Kobe 650, Japan

Receptor-mediated hydrolysis of inositol phospholipids was recently realized to be a common mechanism for transducing various extracellular signals into the cell (2, 7). At an early phase of cellular responses, inositol-1,4,5-trisphosphate (IP_3) mobilizes Ca^{2+}, whereas diacylglycerol activates protein kinase C. These two intracellular mediators are generated from the hydrolysis of phosphatidylinositol-4,5-bisphosphate (PIP_2), and both disappear very quickly. Thus, protein kinase C is active for only a short time after stimulation of the receptor. However, the consequence of protein kinase C activation may persist for a long period, and this enzyme has been proposed to play a role in the expression of plasticity, such as in long-term potentiation of hippocampal neural activity (13). This article will summarize briefly some results of further studies on the structure, intracellular localization, and possible role of protein kinase C in nervous tissues. Several other aspects of this enzyme have also been reviewed recently (4, 5).

I. HETEROGENEITY AND STRUCTURE OF PROTEIN KINASE C

Although protein kinase C preparations from various tissues are apparently similar in their kinetic and catalytic properties, the enzyme isolated from brain soluble fractions appears to be a mixture of several subspecies with slightly different structures. Sequence analysis of the complementary DNAs isolated from bovine (3, 12), rat (5, 11), and rabbit (10) brain complementary DNA libraries has predicted the complete primary sequences of at least three subspecies of the enzyme, which may be derived from different genes. In addition, our recent collaborative studies with the Biotechnology Laboratories, Central Research Division of Takeda Chemical Industries have isolated two types of complimentary DNA clones from rat brain which encode 671 and 673 amino acid sequences. The two subspecies of protein kinase C differ from each other only in the carboxyl terminal regions of about 50 amino acid residues as shown in Fig. 1 (11). This difference appears to be determined by alternative splicing. Although it remains unclear whether several subspecies of protein kinase C are expressed in a single cell or derived from different cell types, it is attractive to imagine that various forms of protein kinase C may transduce different signals into the cell.

II. POSSIBLE ROLES OF PROTEIN KINASE C

During the early phase of cellular responses, IP_3 has been proposed to be an intracellular mediator for Ca^{2+} mobilization from its internal stores as noted above (2). This signal pathway opens transiently, and an increase in the intracellular Ca^{2+} concentration is like a spike in nature. A major mechanism for terminating the signal flow via the IP_3 pathway is thought to be the removal of its 5-phosphate. Similarly, the signal pathway through protein kinase C also opens very rapidly, but normally closes quickly. The diacylglycerol once produced in membranes disappears within a few seconds or, at most, several minutes of its formation. This rapid disappearance of diacylglycerol is due to its conversion back to inositol phospholipids through phosphatidic acid, and to its further degradation to produce arachidonic acid. Although

	Amino acids	Molecular weight
βI	671	76,800
βII	673	76,900
		77,000 (estimated)

Fig. 1. Nucleotide sequences and deduced amino acid sequences of two types of protein kinase C from rat brain, which may be determined by alternative splicing of a messenger RNA derived from a single gene. These enzymes have amino acids of 671 (βI) and 673 (βII), and differ only in the carboxyl terminal region of about 50 amino acids. The molecular weights of these two enzymes well match the value that has been previously estimated by sucrose density gradient analysis. The detailed experimental conditions are described elsewhere (11).

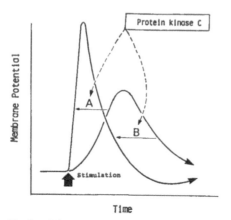

Fig. 2. Schematical representation of modulation of membrane conductance by protein kinase C. A: modulation of ion channels. B: modulation of ion pumps such Ca^{2+}-dependent ATPase.

protein kinase C reveals its activity only for a short time after stimulation of receptors, the consequence of this enzyme activation may persist for a longer period depending upon the biological stability of the phosphate that is covalently attached to each substrate protein molecule. This may provide a biochemical basis for the sustained control of neuronal functions (7), although signal-mediated translocation of this enzyme from the cytosol to the membrane has been proposed to be related to neuronal plasticity, such as hippocampal long-term potentiation (1).

The role of protein kinase C in cell surface signal transduction has been shown for release reactions of neurotransmitters from both peripheral and central nervous systems (7). Potential roles of protein kinase C in stimulus-response coupling may extend to modulation of many other neuronal functions, particularly ion conductance such as channels, pumps, and ion exchanger proteins as schematically shown in Fig. 2 (6, 9). Cytochemical studies with monoclonal antibodies have confirmed that a protein kinase C-positive immunoreactive material is present in many regions of the cytoplasm including dendrites and axons of neuronal cells (7).

III. DOWN REGULATION AND RECEPTOR INTERACTION

In biological systems, positive signals are normally followed by immediate negative feedback control to allow responses to subsequent signals. Although much emphasis has been placed on the positive role of protein kinase C, a major function of this enzyme appears to be related to such feedback control, termed down-regulation, over various steps of its own and other signaling pathways, including the receptors that are coupled to inositol phospholipid hydrolysis and those of some growth factors such as epidermal growth factor and insulin, as schematically shown in Fig. 3. It is worth noting that such dual actions of protein kinase C are attributed entirely to the functional consequence of the phosphorylation of each target protein molecule.

The signaling systems used by cells frequently display extensive heterogeneity, and many variations exist from tissue to tissue. In bidirectional control systems in many tissues, the signals that cause inositol phospholipid hydrolysis promote the activation of cellular functions, whereas the signals that produce cyclic AMP antagonize such activation. For example, in platelets and neutrophils, agonists that elevate cyclic AMP block signal-induced hydrolysis of inositol phospholipids, diacylglycerol formation, and protein kinase C activation as well as cellular responses. This inhibitory action of cyclic AMP extends to the mobilization of Ca^{2+}, presumably through the decreased formation of IP_3 and through the activation of cyclic AMP-dependent protein kinase (protein kinase A). Protein kinase A has a potential to

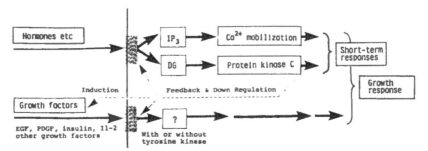

Fig. 3. Schematic representation of signal pathways for short-term and long-term cellular responses. DG, 1,2-*sn*-diacylglycerol.

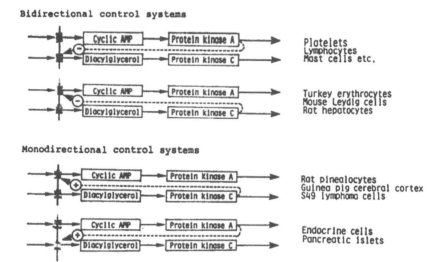

Fig. 4. Interaction of two major signal-transducing systems.

decrease cytosolic Ca^{2+} concentration by phosphorylating the regulatory components of the Ca^{2+}-activated ATPase, thereby enhancing its catalytic activity. In other cell types that appear to be under an inverse form of bidirectional control, such as Leydig cells and ovarian granulosa cells, protein kinase C appears to inhibit and desensitize the adenylate cyclase systems. On the other hand, in several other cell types including pinealocytes and pituitary cells, protein kinase C greatly potentiates cyclic AMP production. In addition, the cyclic AMP-dependent and inositol phospholipid-dependent signal transduction pathways frequently act in concert in many endocrine cells such as pancreatic islets for the release of hormones. These receptor interactions are shown in Fig. 4. The biochemical evidence supporting these interactions is still incomplete, but it is reasonable to assume that various combinations of the two receptor systems may operate positively and be intensified in many neuronal processes (7). Further exploration of various interactions and the network of various signaling systems in individual cell types is of crucial importance for understanding the molecular basis of various neuronal functions.

SUMMARY

The present article summarizes briefly some of our current knowledge of protein kinase C. This research field has seen explosive growth over the past few years, and the evidence available to date seems to indicate the crucial importance of this enzyme in cell-to-cell communication in nervous tissues. Most probably, the protein phosphorylation reactions catalyzed by protein kinase C may exert profound modulation of various Ca^{2+}-mediated processes, including neurotransmitter release, modulation of membrane conductance, down-regulation or feedback control of receptors, interaction with other signaling systems, and many metabolic processes. Further exploration of the role of this protein kinase may provide clues to understand the firm biochemical basis of neuronal plasticity.

Acknowledgment
This work was supported by the Ministry of Education, Science and Culture, Japan and the Muscular Dystrophy Association.

REFERENCES

1 Akers, R.F., Lovinger, D.M., Colley, P.A., Linden, D.J., and Routtenberg, A. (1986). *Science* 231, 587–589.

2 Berridge, M.J. and Irvine, R.F. (1984). *Nature* 312, 315–321.

3 Coussens, L., Parker, P.J., Rhee, L., Yang-Feng, T.L., Chem, E., Waterfield, M.D., Francke, U., and Ullrich, A. (1986). *Science* 233, 859–866.

4 Kikkawa, U. and Nishizuka, Y. (1986). *Annu. Rev. Cell Biol.* 2, 149–178.

5 Knopf, J.L., Lee, M.H., Sulzman, L.A., Kriz, R.W., Loomis, C.R., Hewick, R.M., and Bell, R.M. (1986). *Cell* 46, 491–502.

6 Nishizuka, Y. (1984). *Science* 225, 1365–1370.

7 Nishizuka, Y. (1984). *Nature* 308, 693–698.

8 Nishizuka, Y. (1986). *Science* 233, 305–312.

9 Nishizuka, Y., Kikkawa, U., Kishimoto, A., Nakanishi, H., and Nishiyama, K. (1986). In *Phospholipid Research and the Nervous System*, eds. Horrocks, L.A. and Toffano, L.F., pp. 43–48. Padova: Liviana Press.

10 Ohno, S., Kawasaki, H., Imajoh, S., Suzuki, K., Inagaki, H., Yokokura, H., Sakoh, T., and Hidaka, H. (1987). *Nature* 325, 161–166.

11 Ono, Y., Kurokawa, T., Fujii, T., Kawahara, K., Igarashi, K., Kikkawa, U., Ogita, K., and Nishizuka, Y. (1986). *FEBS Lett.* 206, 347–352.

12 Parker, P.J., Coussens, L., Totty, N., Rhee, L., Young, S., Chen, E., Stable, S., Waterfield, M.D., and Ullrich, A. (1986). *Science* 233, 853–859.

13 Routtenberg, A. (1984). In *Neurobiology of Learning and Memory*, eds. Lynch, G., McGaugh, J., and Weinberger, N., pp. 479–493. New York: Guilford Press.

11

ENHANCEMENT OF CENTRAL SYNAPTIC EFFICACY BY PROLONGED DISUSE OF THE SYNAPSE

MOTOY KUNO

Department of Physiology, Kyoto University, Faculty of Medicine, Kyoto 606, Japan

I. EARLY CONTROVERSY OF DISUSE EFFECTS ON CENTRAL SYNAPSES

Spinal monosynaptic reflexes in the cat show a large and prolonged potentiation following repetitive stimulation of the sensory fibers (*15*). This phenomenon implies that central synaptic function may alter in association with usage. In fact, from a number of early behavioral studies there had been a tacit assumption that use enhances central synaptic function and disuse weakens it. Not until 1951, however, was this assumption experimentally tested. Eccles and McIntyre (*6, 7*) found that several weeks after severing the dorsal roots just distal to their ganglia in the cat, the reflex responses elicited by stimulation of the cut dorsal roots were strikingly reduced, compared with the corresponding reflexes observed on the contralateral, intact side. These results were subsequently confirmed by intracellular recording of monosynaptic excitatory postsynaptic potentials (EPSPs) from spinal motoneurons after section of muscle nerves (*5*). It was then hypothesized that section of the sensory fibers leads to total disuse of the monosynaptic reflex arcs and that synaptic potency is proportional to synaptic use. However, decreased synaptic efficacy might have been due to alterations of the sensory neurons associated with axonal injuries (axon reaction) rather than to the deprivation of impulse activity (disuse). Because of this uncertainty, Beránek and Hník (1) induced another possible disuse condition of the monosynaptic reflex arcs by the severance of muscle tendons (tenotomy). It was assumed that the usage of group Ia sensory fibers arising from muscle spindles is minimized following tenotomy. A few weeks after tenotomy, monosynaptic spinal reflexes evoked by stimulation of the nerve to the tenotomized muscle were found to be significantly enhanced. This suggests that prolonged disuse of the sensory input may cause an increase rather than a decrease in central synaptic efficacy (1). It should be noted, however, that disuse of the sensory pathway from tenotomized muscles has been implied but not experimentally proven. Thus, there has been no definitive evidence as to whether central synaptic efficacy increases or decreases following prolonged disuse of the synapse.

II. TWO FACTORS INVOLVED IN THE MAINTENANCE OF SYNAPTIC EFFICACY

My colleagues and I (9) examined whether the effects of section of the peripheral nerve on central synaptic efficacy can be accounted for entirely by elimination of sensory impulse activity from the muscle. We recorded monosynaptic EPSPs from medial gastrocnemius (MG) motoneurons in response to stimulation of the MG muscle nerve in the hind leg of the cat. In 4 unoperated, control cats, the mean EPSP amplitude was 5.6 mV (Fig. 1A). In 4 other cats, the monosynaptic EPSPs were recorded 2 weeks after section of the MG nerve. In agreement with previous studies (5), the EPSPs were significantly depressed under these conditions, the mean amplitude being 2.7 mV (Fig. 1B). This reduction in the EPSP amplitude could not be attributed to the effects of section of the MG motor axons since monosynaptic EPSPs produced in axotomized MG motoneurons by stimulation of the intact lateral gastrocnemius (LG) nerve remained unchanged (9). Furthermore, monosynaptic EPSPs evoked in intact LG motoneurons by stim-

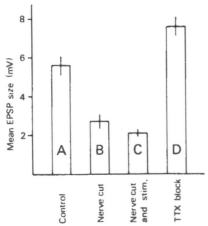

Fig. 1.   Mean amplitudes of monosynaptic EPSPs evoked in MG motoneurons by stimulation of the MG nerve. A: from unoperated, control cats. B: from cats in which the MG
nerve had been cut near the muscle 2 weeks previously. C: from cats in which the sciatic
nerve had chronically been stimulated for 2 weeks after section of the MG nerve. Vertical
bars, S.E.M. cf. ref. 9).

ulation of the cut MG nerve were also depressed (9). Therefore, it was
concluded that reduced EPSP amplitudes were confined only to those
synapses formed by the sensory fibers which had been sectioned.

After section of the MG nerve, impulse activity would no longer
be present in the sensory fibers arising from the MG muscle. If the
observed decrease in the EPSP amplitude were due to prolonged disuse
of the sensory fibers, the reduced synaptic efficacy might be prevented
by daily stimulation of the central stump of the cut MG nerve. To
test this possibility, the sciatic nerve was chronically stimulated (at 100
Hz for 5 sec out of every 30 sec) for 2 weeks after section of the MG
nerve (9). As shown in Fig. 1C, there was no evidence that decreased
synaptic efficacy following section of the MG nerve can be prevented
by added sensory stimuli. In fact, daily stimulation of the sciatic nerve
caused a significant decrease in amplitudes of monosynaptic EPSPs
produced in MG or LG motoneurons by stimulation of the intact LG
nerve.

In order to produce disuse conditions without complications of
nerve injury, impulse conduction of the MG nerve was blocked for 2
weeks in 5 cats, applying a silicone cuff containing tetrodotoxin (TTX)
to the intact MG nerve. From Fig. 1D, it seems clear that monosynaptic

EPSPs evoked by stimulation of the MG nerve are significantly enhanced 2 weeks after conduction block of the MG nerve. Again, the EPSPs elicited in MG motoneurons by stimulation of the intact LG nerve did not differ from those observed in normal cats. Therefore, enhanced synaptic function must result from the deprivation of sensory impulse activity rather than from the possible alterations of MG motoneuron properties by the TTX application to the peripheral nerve.

These results indicate that the effects of severance of a peripheral nerve on central synaptic transmission are entirely different from those observed after elimination of the sensory impulse activity. Therefore, the normal function of central synapses appears to rely on at least two factors: (1) the presence of sensory impulse activity, elimination of which results in synaptic enhancement, and (2) the peripheral axonal continuity of the sensory fibers, interruption of which causes central synaptic depression. Our further studies suggest that the axonal continuity of sensory fibers is essential for the trophic maintenance of function of central synapses formed by the sensory fibers.

III. DOES PERIPHERAL SENSORY REGENERATION RESTORE SYNAPTIC FUNCTION?

If the peripheral axonal continuity of sensory fibers is one factor involved in the maintenance of central synaptic function, central synaptic efficacy once depressed after peripheral nerve injury may be restored following regeneration of the injured sensory fibers. To examine this possibility, we crushed the MG nerve near the muscle and recorded monosynaptic EPSPs from MG motoneurons in response to stimulation of the MG nerve at different post-operative periods (10). Under these conditions, the functional motor reinnervation of the MG muscle was first discernible within 2 weeks after nerve crush and reached its maximal level by the fifth week. As expected, the monosynaptic EPSPs recorded from MG motoneurons were initially depressed after nerve crush (Fig. 2). By the end of the fourth or fifth week, the EPSPs had recovered to the near-normal level. This period was then followed by a further increase in the EPSP amplitude before subsiding to normal levels (Fig. 2). Essentially the same profile was observed for the am-

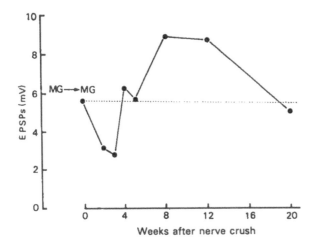

Fig. 2. Changes in the mean amplitudes of monosynaptic EPSPs evoked in MG moto-neurons by stimulation of the MG nerve after crushing of the nerve in the cat. (cf. ref. 10).

plitude changes of monosynaptic EPSPs evoked in intact LG moto-neurons by stimulation of the MG nerve (10).

While central synaptic efficacy once depressed after peripheral nerve crush was undoubtedly restored in association with peripheral regeneration, the supernormal enhancement of synaptic efficacy oc-curring 8–12 weeks after nerve crush was rather puzzling. During the supernormal period (week 8), some MG sensory fibers showed func-tional recovery, whereas others failed to respond to muscle stretch. We examined whether the functionally regenerated MG sensory fibers may produce significantly larger EPSPs than normal MG sensory fibers. Figure 3 shows examples of monosynaptic EPSPs recorded from MG motoneurons in response to impulses in a single sensory fiber elicited by stretch of the MG muscle in a control, unoperated cat (left records) and in a cat whose MG nerve was crushed 8 weeks previously (right records), using the spike-triggered averaging method (16). Single fiber-EPSPs were recorded, in total, from 79 MG motoneurons in control animals and from 100 MG motoneurons in nerve-crushed animals. The mean amplitudes (220 μV and 203 μV) of single fiber-EPSPs were in-distinguishable between the control and operated animals. Therefore, it is suggested that the sensory fibers responsible for the supernormal

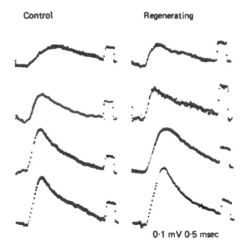

Control   Regenerating

0·1 mV 0·5 msec

Fig. 3. Monosynaptic EPSPs evoked in different MG motoneurons in response to impulses in a single MG sensory fiber elicited by stretch of the MG muscle. Left records, from an unoperated, control cat. Right records, from a cat whose MG nerve had been crushed 8 weeks previously. Calibration at the end of each record. (*cf. ref. 10*).

phase of the composite EPSPs (Fig. 2) are those which regenerate into the muscle but fail to form functional sensory reinnervation. This condition may be considered analogous to the situation of normal sensory fibers whose conduction is chronically blocked with TTX, thus preserving the peripheral axonal continuity but halting impulse activity. Under such conditions, the EPSP amplitude is expected to be enhanced (Fig. 1D). Thus, alterations of monosynaptic EPSPs during peripheral regeneration of the crushed sensory fibers are consistent with the presence of two distinct factors involved in the maintenance of normal function of central synapses.

IV. WHAT IS REQUIRED FOR RESTORATION OF CENTRAL SYNAPTIC FUNCTION?

If the axonal continuity of sensory fibers is essential for the maintenance of central synaptic function, the synaptic responses may eventually disappear if regeneration of the cut MG nerve is prevented for long periods. To test this question, the central stump of the cut MG nerve was ligated, deflected centrally and anchored to the intact semitendinosus muscle to impede its regeneration into the MG muscle for periods up to 8 months (12). The EPSPs virtually disappeared 6 months ($<10\%$ of the normal size) or 8 months ($<5\%$ of the normal size) after nerve section. We then tested whether the severely depressed synaptic responses can still recover following restoration of peripheral sensory connections. For this purpose, the cut MG nerve was reunited with its muscle following a post-denervation delay of 6 months, and the EPSPs were recorded 2 months later. After a 6-month denervation period, the peripheral reunion practically failed to form any functional motor connections. Also, the wet weight of the reinnervated MG muscle was only 20% of that of the control, contralateral MG muscle. Furthermore, only 2 out of 51 sensory fibers examined showed normal responses to stretch of the reinnervated MG muscle under these conditions. However, the mean amplitudes of monosynaptic EPSPs evoked in MG motoneurons (open circles) or LG motoneurons (filled circles) by stimulation of the MG nerve were not significantly different from the normal value (Fig. 4b). Also, these values were practically the same as those obtained 2 months after immediate reunion of the cut MG nerve

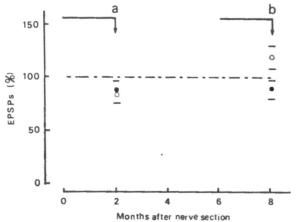

Fig. 4. The mean amplitudes of monosynaptic EPSPs evoked in MG (O) and LG (●) motoneurons by stimulation of the MG nerve. a: from cats in which the EPSPs were recorded 2 months after immediate reunion of the cut MG nerve. b: from cats in which the EPSPs were recorded 2 months after reunion of the cut MG nerve following a post-denervation delay of 6 months. Horizontal bars on top indicate a 2-month period following reunion of the cut MG nerve. Arrows indicate the time of the EPSP observation at the end of the 2-month period. Bars near each point, S.E.M. Each point is plotted as a percentage of the control EPSP amplitude observed in unoperated, control cats. (cf. ref. 12).

(Fig. 4a). Therefore, it is concluded that recovery of central synaptic function following regeneration of the cut peripheral nerve does not require recovery of muscle activity nor restoration of functional sensory activity. On the other hand, the EPSP decrement proceeded as long as peripheral regeneration was prevented. Thus, the recovery of central synaptic function must be triggered by some factors associated with reconnection of the cut peripheral nerve to the muscle. It should be noted that during prevention of regeneration, the central stump of the cut nerve was united with the intact semitendinosus muscle. This suggests that recovery of central synaptic function requires mechanical contacts of the cut peripheral nerve to some denervated muscle. The most likely possibility would be then that the recovery of central synaptic function may be triggered by some trophic factor supplied by contact of the peripheral sensory terminals with the denervated muscle.

V. EFFECTS OF NERVE GROWTH FACTOR ON CENTRAL SYNAPTIC FUNCTION

Recently, nerve growth factor (NGF) has been shown to play a trophic role on sensory neurons early in postnatal or even adult animals. Thus, the substance P content in sensory neurons decreases following chronic section of the peripheral nerve, but this reduction can be prevented by daily administration of NGF (8, 11, 18). Similarly, fluoride-resistant acid phosphatase in the dorsal horn, which is normally depleted after peripheral nerve section, can be maintained by NGF treatment (4, 8). We have examined whether depression of monosynaptic EPSPs produced by peripheral nerve crush may be affected by daily treatment with NGF in neonatal rats (13, 17). For this purpose, the MG nerve was crushed near the muscle in Wistar rats on the day after birth. The monosynaptic EPSPs evoked by stimulation of the MG nerve were then recorded after a post-operative period of 24–35 days with or without daily injections of 2.5S NGF purified from the mouse submaxillary glands (2). Figure 5 shows the mean amplitudes of monosynaptic EPSPs elicited in MG motoneurons by stimulation of the MG nerve. In 20 age-matched normal rats, the mean EPSP amplitude was 2.4 mV (A). In 8 rats in which the MG nerve was crushed on the day after birth, the EPSPs were strikingly reduced, the mean amplitude being only 0.31

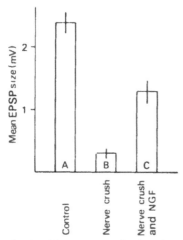

Fig. 5. The mean amplitudes of monosynaptic EPSPs evoked in MG motoneurons by stimulation of the MG nerve in rats, 25–36 days in age. A: from unoperated, control rats. B: from rats whose MG nerves had been crushed on the day after birth near the muscle. C: same as in B, but the animal was daily treated with NGF. Vertical bars, S.E.M. (*cf.* ref. *17*).

mV (B: 13% of the control size). In 11 animals, the same operation was performed, but the rats were daily treated with NGF (2 μg/g body weight) from the day of the operation. In these animals, the EPSPs (C: 1.3 mV; 54% of the control size) were 4.2 times greater than those observed after nerve crush alone. Daily administration of NGF also increased monosynaptic EPSPs evoked by stimulation of the intact LG muscle nerve. However, this increment (about 30%) was marginal compared with increased EPSPs in response to stimulation of the crushed MG nerve (about 400%). Thus, enhancement of the EPSPs by NGF depends upon the state of the sensory neurons.

It might be argued that peripheral nerve crush in neonatal rats may cause cell death in some of the sensory neurons and that daily treatment with NGF may have a remedial effect on cell death. This may account for the greater enhancement of EPSPs evoked by stimulation of the crushed nerve than those produced by stimulation of the intact nerve. To test this possibility, all the MG sensory neurons in the lumbar dorsal root ganglia were counted by labeling with horseradish peroxidase (HRP). On the average, nerve crush caused a loss of about 45% of the MG sensory neurons. The cell death of small sensory neurons (presumably $A\delta$ and C cells) was prevented by daily treatment with NGF, whereas the NGF treatment was ineffective in preventing the cell death of large ($A\alpha$) sensory neurons. Since the sensory neurons responsible for the generation of monosynaptic EPSPs are large group Ia neurons, enhancement of the EPSPs evoked by stimulation of the crushed MG nerve following NGF treatment cannot be attributed to the rescue of axotomized sensory neurons that are destined to die. Also, if decreased EPSP amplitude following nerve crush were entirely due to the loss of sensory neurons, the EPSP amplitude should be about 55% of the control size, in contrast with 13% (Fig. 5B). Therefore, synaptic depression must occur at those synapses formed by the surviving sensory neurons whose axons had been crushed. NGF treatment restored the EPSP amplitude to about 54% of the control size (Fig. 5C). Thus, an increase in the EPSP amplitude by NGF can be attributed to an almost full reversal of the depression at the synapses formed by axotomized sensory fibers. At present, it is not certain whether the central synaptic function of group Ia sensory neurons is normally maintained by endogenous NGF. However, these results do show that central synaptic function of Ia sensory neurons is responsive to NGF in postnatal life when the peripheral axonal continuity is interrupted. This strongly suggests the involvement of some trophic factor in the maintenance of central synaptic function. The responsiveness of group Ia sensory neurons to NGF in neonatal rats has also been supported by the formation of *de novo* muscle spindles in temporarily denervated muscles following treatment with NGF (*19*).

VI. DOES IMPULSE ELIMINATION BY NERVE SECTION ENHANCE SYNAPTIC FUNCTION?

It now seems clear that the loss of sensory impulse activity leads to enhancement of the central synaptic function. The sensory impulse activity is also eliminated by section of a peripheral nerve. However, one week after nerve section, the amplitude of monosynaptic EPSPs evoked by stimulation of the cut nerve did not differ from the normal EPSP amplitude (*9*). Why does nerve section produce no enhancement of the central synaptic function? This question has recently been examined by Y. Miyata and H. Yasuda in adult rats. They have found that monosynaptic EPSPs evoked by stimulation of the MG nerve begin to increase one day after section of the MG nerve, reaching the maximal level on the third post-operative day (Y. Miyata and H. Yasuda, personal communication). Furthermore, the amplitude of these EPSPs declined to near-normal levels one week after nerve section before undergoing synaptic depression. Evidently, this initial supernormal phase has been overlooked in early axotomy studies on the cat. It is almost certain that this initial enhancement is due to the elimination of sensory impulse activity by nerve section, whereas the subsequent depression is induced by interruption of the peripheral axonal continuity. Thus, the effects exerted by the loss of the two factors on central synaptic function appear to have different time courses.

How the deprivation of sensory impulse activity increases central synaptic function is a matter of speculation at present. Lasek and Hoffman (14) have suggested that terminal sprouting is normally prevented by the continual disassembly of the axonal cytoskeleton owing to the entry of calcium in the nerve terminal. With TTX blocking impulse activity, the amount of calcium entering the terminal normally associated with excitation would be reduced. Under such conditions, central synaptic efficacy may be increased as a result of sprouting of central terminals of the sensory neurons. The presence of this mechanism has recently been shown at neuromuscular junctions (3).

SUMMARY

There has been a controversy as to whether the efficacy of central synapses increases or decreases following prolonged disuse of the synapses. We have examined this problem, using monosynaptic EPSPs evoked in spinal motoneurons by stimulation of the MG nerve as criteria. When the MG nerve of the cat was cut, the EPSPs produced by stimulation of the MG nerve were depressed within 2 weeks. Decreased synaptic efficacy could not be prevented by daily stimulation of the central stump of the cut MG nerve. Also, when the intact MG nerve was chronically blocked with TTX for 2 weeks, the EPSPs elicited by stimulation of the MG nerve became greater than the normal EPSPs. Thus, the normal central synaptic function appears to rely on at least two factors: (1) the presence of sensory impulse activity, elimination of which results in synaptic enhancement, and (2) the peripheral axonal continuity of the sensory fibers, interruption of which causes synaptic depression. When the MG nerve was cut and prevented from reinnervation for 6 months, the EPSPs in response to stimulation of the MG nerve were virtually abolished. However, if the cut MG nerve was rejoined to the muscle following a post-denervation delay of 6 months, the EPSPs were restored to near-normal levels. This restoration did not require muscle activity nor sensory impulse activity. It is suggested that the recovery of central synaptic function is triggered by some trophic factor supplied by contact of the peripheral sensory terminals with the denervated muscle. Daily treatment with nerve growth factor after section of the MG nerve in neonatal rats reversed central synaptic depression produced by section of the peripheral nerve. It is concluded that the normal function of central synapses is maintained by a retrograde trophic factor through the sensory fibers in addition to neural activity of the sensory fibers.

REFERENCES

1 Beránek, R. and Hnik, P. (1959). *Science* **130**, 981–982.
2 Bocchini, V. and Angeletti, P.U. (1969). *Proc. Natl. Acad. Sci. U.S.A.* **64** 787–794.
3 Connold, A.L., Evers, J.V., and Vrbová, G. (1986). *Dev. Brain Res.* **28**, 99–107.
4 Csillik, B., Schwab, M.E., and Thoenen, H. (1985). *Brain Res.* **331**, 11–15.
5 Eccles, J.C., Krnjević, K., and Miledi, R. (1959). *J. Physiol.* **145**, 204–220.
6 Eccles, J.C. and McIntyre, A.K. (1951). *Nature* **167**, 466–468.
7 Eccles, J.C. and McIntyre, A.K. (1953). *J. Physiol.* **121**, 492–516.
8 Fitzgerald, M., Wall, P.D., Goedert, M., and Emson, P.C. (1985). *Brain Res.* **332**, 131–141.
9 Gallego, R., Kuno, M., Núñez, R., and Snider, W.D. (1979). *J. Physiol.* **291**, 191–205.
10 Gallego, R., Kuno, M., Núñez, R., and Snider, W.D. (1980). *J. Physiol.* **306**, 205–218.
11 Goedert, M., Stoeckel, K., and Otten, U. (1981). *Proc. Natl. Acad. Sci. U.S.A.* **78**, 5895–5898.
12 Goldring, J., Kuno, M., Núñez, R., and Snider, W.D. (1980). *J. Physiol.* **309**, 185–198.
13 Kuno, M., Miyata, Y., Homma, S., and Ogawa, M. (1985). *Neurosci. Res.* **2**, 275–280.
14 Lasek, R.J. and Hoffman, P.N. (1976). In *Cell Motility*, eds. Pollard, T. and Rosenbaum, J., pp. 1021–1049. New York: Cold Spring Harbor.
15 Lloyd, D.P.C. (1949). *J. Gen. Physiol.* **33**, 147–170.
16 Mendell, L.M. and Henneman, E. (1971). *J. Neurophysiol.* **34**, 171–187.
17 Miyata, Y., Kashihara, Y., Homma, S., and Kuno, M. (1986). *J. Neurosci.* **6**, 2012–2018.
18 Otten, U., Goedert, M., Mayer, N., and Lembeck, F. (1980). *Nature* **287**, 158–159.
19 Sekiya, S., Homma, S., Miyata, Y., and Kuno, M. (1986). *J. Neurosci.* **6**, 2019–2025.

OPINIONS ON STRATEGY
AND PROSPECTS IN NEUROSCIENCE

SOME THOUGHTS ON REGENERATION

ERIC M. SHOOTER

Department of Neurobiology, Stanford University School of Medicine, Stanford, CA 94305, U.S.A.

The recent development of new techniques in anatomy, physiology, and molecular genetics provides us with unrivalled opportunities to increase our understanding of how the nervous system develops, functions, and dies. In this short article I will focus on a question of both fundamental and practical importance, namely the problem of neuronal regeneration, because it illustrates both the potential and the power of some of this new methodology. The inability of most neurons to divide once they have differentiated imposes on the nervous system unique problems of survival in the face of a variety of environmental insults and traumatic injury. It is becoming clear that the nervous system must rely on a complex series of repair mechanisms in order to continue to function properly over time periods often measured in decades. A reasonable analogy can be made to the repair mechanisms involved in maintaining DNA structure and thus the integrity of the genome. While these repair mechanisms are most obviously considered in the context of traumatic injury they must also function throughout the lifetime of individual differentiated neurons to maintain their viability. Moreover, many of these mechanisms operate during the development of the nervous system controlling not only growth but selective survival, and selective modifications of neurons. It is convenient to divide the mechanisms into those which operate within neurons (intrinsic) and those which influence neuronal growth properties from the external milieu (extrinsic). The two are related in that, in the context of growth, the former are the targets of the latter.

Intrinsic mechanisms controlling axonal growth
In the same way that the activity of a single oncogene can lead to uncontrolled growth of a cell it is reasonable to assume that the activity

of one or a few genes will control axonal growth. By looking to see which neuronal proteins were induced after nerve injury it was possible to identify, besides the major cytoskeletal proteins, a small set of proteins whose abundant expression always accompanied successful axon regeneration (13). The synthesis, for example, of one of these so-called growth-associated proteins, GAP-43, was high in regenerating or developing neurons but not in non-regenerating central nervous system (CNS) nerves after injury. Further evidence for the putative role of GAP-43 in axonal growth comes from the finding that it is inserted into the growth cone membranes, that it is phylogenetically conserved and that it is a target for phosphorylation by protein kinase C (14). However, the proof that GAP-43, or other proteins, are regulators of axonal growth requires the ability to modulate levels of these proteins in appropriate systems and observe the accompanying morphological changes. This can now be achieved, in principle, in both negative and positive directions. The cloning of the cDNA for GAP-43 will allow the synthesis in one of several vectors of both sense and antisense mRNAs. The latter can be injected into PC12 pheochromocytoma cells to determine if, in inhibiting expression of endogenous mRNA for GAP-43, it also prevents nerve growth factor (NGF)-induced neurite outgrowth. The converse of this experiment would be to see if expression of the sense mRNA for GAP-43 in undifferentiated cells leads to neurite outgrowth in the absence of NGF. Variations of these experiments would be the transfection of PC12 cells with vectors containing an inducible promoter, such as the metallothionein promoter, and a gene for dehydrofolate reductase to permit amplification, as well as the GAP-43 cDNA in one or other orientation. The extension of these procedures to introduce an inducible GAP-43 gene into transgenic animals (4) is also now feasible.

How many genes are involved? Judging by the number of proteins which meet the criteria of (1) being significantly induced in regenerating but not non-regenerating neurons, (2) being widely expressed in developing neurons, both peripheral nervous system (PNS) and CNS, and (3) being phylogenetically conserved, the answer is probably very few. However, the screen for these proteins so far has been at the level of protein synthesis. A more definitive screen will use cDNA libraries from an appropriate population of neurons undergoing regeneration.

Such libraries can be screened with probes enriched for putative GAP cDNAs. The latter are prepared by hybridizing the single-stranded cDNA from the regenerating neurons with polyA+ mRNA from the same neurons in the mature non-growth state, thus removing all the common constitutive cDNAs and leaving only the cDNAs for the GAPs. This method has been used successfully to isolate the cDNA for the rat NGF receptor without relying on any knowledge of its structure (*10*).

Extrinsic mechanisms controlling axonal growth
That such mechanisms exist is known from a variety of approaches, perhaps none more direct than the observation that mammalian CNS neurons, which normally do not regenerate, will regrow axons in a peripheral nerve graft (*15*). The most obvious candidates are the trophic factors such as NGF. Not only do the targets of sympathetic and sensory neurons provide a source of NGF for axonal regrowth but so do the proliferating non-neuronal cells in the nerve distal to a site of injury (*6*). Since the Schwann cell population also expresses NGF receptors (*16*) the growth cones of the regenerating axons move over layers of NGF in the distal nerve stump permitting close interaction between this layer and NGF receptors on the growth cone. The factors involved in the regulation of NGF transcription in the target and in non-neuronal cells and of NGF receptors on the Schwann cells are unknown, other than that they are of neuronal origin, but the tools, the characterized genes for NGF and for its receptor, are available to tackle this problem. It will be of great interest to learn whether the regulation is due to repression or activation of the genes involved.

The recent observations that the cholinergic neurons of the basal forebrain nuclei accumulate NGF *via* retrograde transport in their hippocampal and cortical projections (*12*) identifies NGF as a trophic factor for one population of neurons in the CNS and immediately focuses attention on the possible role of growth factors in the etiology of CNS neurological disease (*5*). All this raises the question of how many "NGF" exist and how can they be isolated? Two approaches have been successful and will no doubt be used in the future. In the first, highly selective and efficient protein purification procedures have been used to isolate minute amounts of a brain derived neurotrophic factor (BDNF) where biological activity was defined by its ability to promote the survival of a well defined population of neurons. It can be estimated from the known NGF concentration in sympathetic targets that purification factors of at least one million are required to achieve final purification of growth factors in the same class as NGF. The spectrum of activity of BDNF differs from that of NGF. Sensory neurons in dorsal root and nodose ganglia and retinal ganglion cells but not sympathetic neurons, are the targets of BDNF (1, 17) and in particular, given its origin, BDNF supports the central projections of sensory neurons. This work is the prototype of what will be needed to isolate other factors. Given sufficient amounts of the protein to establish partial amino acid sequences, oligonucleotide probes can be synthesized to correspond to these sequences and used, in turn, to isolate the cDNA and gene for the protein. The expression of either in currently available expression systems should provide adequate quantities of the protein to fully define its biological role. In the second approach, growth factors isolated for their known mitogenic activity on non-neuronal cells have been re-examined for their potential neurotrophic activity on brain neurons. Thus both acidic and basic fibroblast growth factor support the survival of hippocampal neurons while the basic factor also markedly enhances neurite outgrowth from these neurons (18). Epidermal growth factor has similar effects on neocortical neurons (9). Clearly the activities of other proto-oncogene products should be examined for their putative neurotrophic activities. From available data it seems likely that only a relatively small number of target-derived growth factors will be needed to support the major neuronal groups in the nervous system.

The second class of neurotrophic agents are proteins associated with the extracellular matrix (17) and with cell surfaces. Besides their role in controlling axonal cell growth it is likely that these molecules are important also in axonal guidance. The extracellular matrix proteins were first implicated by observations that non-neuronal cells secrete neurotrophic molecules which bound to the matrix. Laminin is one such protein and it induces neurite outgrowth in a variety of neuronal types. However, other cell surface molecules which are not matrix proteins also have this property, as demonstrated by the fact that glial and muscle cell surfaces are excellent substrates for axonal growth. The identification of these molecules and perhaps other matrix proteins is made possible by the monoclonal antibody methodology. Initially one would test monoclonal antibodies against known matrix proteins for their ability to block neurite growth on glial or muscle surfaces. The anticipation is that these antibodies would be ineffective because it is known, for example, that retinal ganglion cells that lose their ability to respond to laminin still grow neurites on astrocytes (2). Monoclonal antibodies would then be sought which block neurite outgrowth from various neurons and through the use of specific antibody affinity columns it should be possible to isolate the proteins involved. It will be of considerable interest to see if any of these proteins are homologous to the class of cell adhesion proteins exemplified by neural cell adhesive molecule (N-CAM) (3). This strategy can be extended to characterize and isolate glial surface proteins which are known to inhibit neurite outgrowth (11) by asking if any of the monoclonal antibodies, by reacting with such proteins, now allow neurite outgrowth on the inhibitory cells. Both extracellular matrix proteins, such as laminin, and the cell surface proteins exert their effects through specific cell surface receptors on the neuronal membrane. The characterization of these receptors, probably by gene transfer and cloning as used for the NGF receptor, and of the signal transducing mechanisms will probably follow quite rapidly.

Strategies for identifying other extrinsic mechanisms
The two classes of neurotrophic agents discussed above, target-derived factors and extracellular matrix or cell surface molecules do not cover all the possibilities for agents induced in non-neuronal cells of the injured nerve. The most effective procedure for the identification of other proteins controlling axonal growth is that of differential hybridization of the cDNAs and mRNAs from control and regenerating nerve. For example, cDNA libaries can be made by standard procedures in vectors such as pUC9, from the total mRNA obtained from control and injured rat sciatic nerve at different times after injury. These may be referred to as the—(control) and + (regenerating) libraries, respectively. Single stranded cDNA probes are made from the same total mRNA and identified as − and + probes, as defined above, and then used to screen the libraries. A comparison of the hybridization

of the + library with − and + probes, for example, will identify genes induced by nerve injury while a similar screening of the − library will identify genes repressed by injury. The amount of information extracted by this procedure depends on the stringency of the hybridization. In pilot experiments using a different model system, *i.e.*, the differentiation of PC12 cells by NGF, three clones were identified from an initial screen of several hundred clones as representing cDNA sequences induced by NGF. Analysis of the mRNA from differentiated compared to undifferentiated PC12 cells confirmed this point. The cDNAs were sequenced to derive the structure of the corresponding proteins. Two of these proteins turn out to be homologous, but not identical, to the S-100, calcium binding proteins. The interesting point here is that another protein of this class has been recently identified as a neurite-promoting factor (*8*). Clearly it is impossible to predict, at this stage, how many cDNAs will be identified and the magnitude of the task involved in identifying the corresponding proteins but the example given above suggests that some of them will be important to the regulation of axonal growth.

Apolipoproteins in nerve degeneration and regeneration

Another example suffices to show that it is well worthwhile following up changes in gene expression after nerve injury to explore the molecular basis of successful nerve regeneration. In a survey of changes in levels of protein synthesis after rat sciatic nerve injury one protein was identified whose synthesis increased markedly in the distal stump (7). Cells in the nerve sheath produce and secrete relatively large amounts of this protein in the several weeks after injury. However, by the time the nerve fibers have regrown and the process of remyelination is significantly advanced, both levels of synthesis and of accumulation of this protein are back down to the very low levels found in mature, intact nerve. This protein has been identified as apolipoprotein E (apo E). Apo E is a well-studied serum protein that is involved in lipid and cholesterol metabolism and it is likely to have similar functions in the nervous system. It has been known for a long time that lipid from degenerating peripheral axons and myelin is stored locally in Schwann cells and macrophage and that it can be reutilized, at least in remyelination (Fig. 1). The presence of apo E in the injured

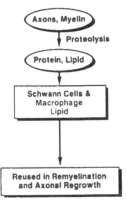

Fig. 1. Lipid metabolism after nerve injury: the fate of lipid in peripheral nerve after injury.

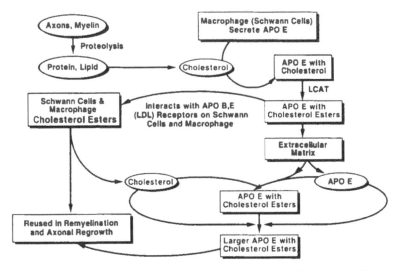

Fig. 2. Lipid metabolism after nerve injury. A model describing how apo E can shuttle lipid (cholesterol is used as the example) from degenerating nerve to Schwann cells and macrophage for storage and then between Schwann cell, macrophage and the growth cone in regenerating nerve for remyelination and axonal regrowth.

nerve provides the possible molecular mechanism behind these events (Fig. 2). The apo E binds the released myelin and axonal lipid and through interaction with apo E receptors on Schwann cells and macrophage carries the lipid into these cells for storage. The apo E-lipid particles also bind to the extracellular matrix along the path of the degenerating nerve. By these processes little lipid is lost and indeed is available, by reversal of the above cycle, for renewed axonal growth (where possible) and remyelination. In the CNS apo E production is also initiated after nerve injury, but in astrocytes rather than macrophage. Interestingly, little if any storage of apo E-bound lipid ensues either in macrophage, astrocytes or extracellular matrix.

In this instance the search for a neurotrophic factor identified a fundamental mechanism of lipid metabolism in the nervous system and brought out major differences between PNS and CNS in the way lipid metabolism is handled. Do these differences account, in part, for the much slower degeneration of axonal and myelin lipid in the CNS and what role does the lack of accumulation of apo E in the CNS play, if any, in its inability to regenerate? Again, the availability of cDNA probes for apo E and the apo E receptor will make it possible to determine the signals which turn on or off apo E and apo E receptor gene transcription in macrophage and astrocytes.

Prospects

The examples given above of the application of molecular genetics to a discrete problem in neurobiology are obviously reflected in other areas of great interest, such as ion channel structure and function, neuropeptide processing, and signal transduction in relation to memory, to name just a few. That very significant new insights will accrue over the next few years is obvious. Returning to the field of regeneration for one last comment in this regard, it is clear that it will benefit from the major emphasis on transplantation. The finding that neurons in some transplants can reestablish contact with appropriate target fields even though their axons do not traverse the usual pathways suggests that neurons can retain their specific surface molecular clues for some time after injury. Whether this will ultimately lead to the reestablishment of the fine specificity of connections remains to be seen.

REFERENCES

1 Barde, Y.A., Edgar, D., and Thoenen, H. (1983). *Annu. Rev. Physiol.* **45**, 601.
2 Cohen, J., Burne, J.F., Winter, J., and Bartlett, P. (1986). *Nature* **322**, 465.
3 Edelman, G. (1985). *Annu. Rev. Biochem.* **54**, 135.
4 Gordon, J.N., Scango, G.A., Plotkin, D.J., Barbosa, J.A., and Ruddle, F.H. (1980).

Proc. Natl. Acad. Sci. U.S.A. **77**, 7380.

5 Hefti, F. and Weiner, W.J. (1986). *Ann. Neurol.* **20**, 275.

6 Heumann, R., Korshing, S., Bandtlow, C., and Thoenen, H. (1987). *J. Cell Biol.* **104**, 1623.

7 Ignatius, M.J., Gebicke-Härter, P.J., Skene, J.H.P., Schilling, J.W., Weisgraber, K.H., Mahley, R.W., and Shooter, E.M. (1986). *Proc. Natl. Acad. Sci. U.S.A.* **83**, 1125.

8 Kligman, D.K. and Marshak, D.R. (1985). *Proc. Natl. Acad. Sci. U.S.A.* **82**, 7136.

9 Kornblum, H.L., Morrison, R.S., Bradshaw, R.A., and Leslie, F.M. (1986). *Soc. Neurosci. Abstr.* **12**, 1101.

10 Radeke, M.J., Misko, T.P., Hsu, C., Herzenberg, L.A., and Shooter, E.M. (1987). *Nature* **325**, 593.

11 Schwab, M.E. and Thoenen, H. (1985). *J. Neurosci.* **5**, 2415.

12 Seiler, M. and Schwab, M.E. (1984). *Brain Res.* **300**, 33.

13 Skene, J.H.P. (1984). *Cell* **37**, 697.

14 Skene, J.H.P., Jacobson, R.D., Snipes, G.J., McGuire, C.B., Norden, J.J., and Freeman, J.A. (1986). *Science* **233**, 783.

15 So, K-F. and Aguayo, A.J. (1985). *Brain Res.* **328**, 349.

16 Taniuchi, M., Clark, H.B., and Johnson, E.M., Jr. (1986). *Proc. Natl. Acad. Sci. U.S.A.* **83**, 4094.

17 Thoenen, H. and Edgar, D. (1985). *Science* **229**, 238.

18 Walicke, P. and Cowan, W.M. (1986). *Soc. Neurosci. Abstr.* **12**, 1101.

COMMENT ON THE RELATIONSHIP BETWEEN CLINICAL RESEARCH AND NEUROSCIENCE, AND ON A STRANGE COMPOUND IN IMMATURE BRAINS

YASUO KAKIMOTO

Department of Neuropsychiatry, Ehime University School of Medicine, Onsengun, Ehime 791-02, Japan

An old Japanese proverb is: "Furuki o tazune atarashiki o shiru," which means that we can see the future by looking at history. This is appropriate for this symposium held in the old Japanese capital of Kyoto.

I would like to comment on two points. One concerns the relationship between clinical research and neuroscience. Clinical research and experience have been important skeletons of neuroscience in the past and will be in the future. This is also true in other fields of biological science, and it is especially true in neuroscience since most of the brain functions are expressed as human behavior such as social functions, disturbances in the learning and communication of which are expressed as psychiatric or neurological disorders. The second topic concerns a compound in the fetal or neonatal brain found in our laboratory; this is thought to be related to cell differentiation.

For the first subject, I would like to speak about Parkinson's disease. This was the entity of a disease written of by Dr. Parkinson in 1817 (12) in his essay on shaking palsy. He described its clinical symptoms almost completely. Charcot later called it Parkinson's disease, and neuropathological studies revealed the pathological changes in substantia nigra by Brissaund, and then in basal ganglia around the beginning of this century. Dopamine (DA) was found in the brain by Carlsson et al. (3) in 1958 and its distribution in the brain was investigated independently by the Bertler's (2) and Sano's groups (13). I was a graduate student working in Dr. Sano's laboratory, and I developed a method of determination of DA and participated in this work. I remember well how surprised we were to find the striking localization of DA in the basal ganglia; this was unexpected since DA had been considered to function only as an intermediate of norepinephrine synthesis. The next year Sano's group (14) and Ehringer's group (4) independently discovered the decrease in DA concentration in the nigrostriatal system of the autopsied brains of patients with Parkinson's disease.

Trials of the use of L-DOPA for the treatment of Parkinson's disease started (1) and it was found to be effective. It is amazing that as long as 170 years ago Parkinson described the almost complete frame of clinical pictures of the disease on which L-DOPA is effective. DOPA therapy was a monument created by the cooperative work of neurology, pathology, and neurochemistry.

Neurologists had been optimistic for the first few years following the discovery of DOPA therapy, but they soon realized that L-DOPA was not effective in preventing the progress of neurological disturbance of Parkinson's disease. In addition, about half of the patients showed hallucination or delirium during DOPA therapy. My colleagues and I (5) treated about 200 patients and found that those exhibiting these psychotic symptoms had dementia. Slow EEG and cerebral cortical atrophy were also present in demented patients. In the recent several years there have been many reports of clinical, pathological and chemical findings in patients with Parkinson's disease similar to those found in Alzheimer's disease. Recent studies on methylphenyltetrahydropyridine (MPTP) intoxication, genetic studies, and recent biotechnological findings on neuronal degeneration promise a second stage of research development for elucidation of the true etiology and therapy of Parkinson's disease.

Next, a comment about psychopharmacology. Laborit (6) found chlorpromazine useful for the treatment of shock, and Delay (7) happened to use chlorpromazine on psychotic patients which revealed that the drug was especially effective in inhibiting delusion, hallucination and psychomotor excitation. This triggered a revolutional change in the treatment of psychosis. Many derivatives, butyropherons and many other drugs were developed. Carlsson's work, which showed an increase in the number of DA metabolites by chlorpromazine in the

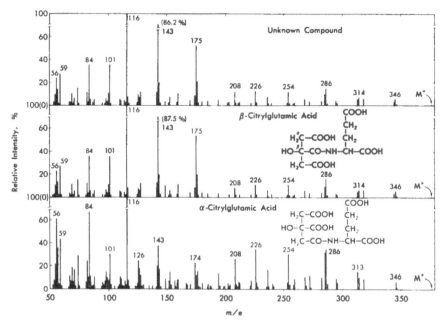

Fig. 1. Mass spectra of the methyl esterified unknown compound and the synthesized tetramethyl esters of α- and β-citrylglutamic acid. The spectra of the methyl esterified unknown compound and the tetramethyl esters of α- and β-citrylglutamic acid were recorded at inlet temperature 40°C and ionization energy of 70 eV by a Nihon Denshi, JEOL JMS-D-300 mass spectrometer.

brain of rat treated with monoamine oxidase inhibitor, provided a clue to the mechanism of the drug's action. Carlsson's work was followed by determination of DA turnover and binding of neuroleptic drugs with DA receptors. It is agreed that the blocking of DA receptor in the brain is closely related to the pharmacological actions of neuroleptics. This also explains the side effect of neuroleptics, Parkinsonism.

Many groups of drugs have been developed through clinical and cultural experiences, but not on the theoretical bases of neuroscience. Neuroleptics, antidepressants, lithium, benzodiazepines, antiepileptics, hypnotics, and analgetics fall into this category. Studies on the mechanism of action have recently been greatly accelerated by the use of sophisticated techniques of biotechnology. Neuroscience is an integrated field of many disciplines, and incorporation of recent knowledge and techniques of molecular biology is undoubtedly stimulating its rate of progress. I would like to stress again that clinical research and experi-

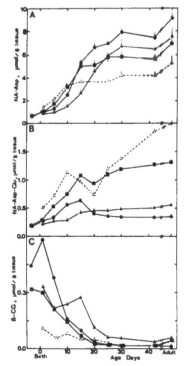

Fig. 2. Changes of NA-Asp (A), NA-Asp-Glu (B) and β-CG (C) during development of the rat. For each determination up to 10 days of age tissues were pooled from two to four animals. NA-Asp, NA-Asp-Glu, and β-CG were determined by gas chromatography after extraction with 75% ethanol. Values were corrected on the basis of the recovery rate and expressed as means of four to six samples (the vertical bars represent S.D.). ● cerebrum; ▲ cerebellum; ■ brainstem; ▲ spinal cord.

ence are still the essential skeleton of neuroscience. Clinical knowledge will become even greater with newly developed methods using computer technology and in the light of new information on neuroscience. Thus, neuropsychiatry and neuroscience will progress hand in hand.

With reference to the second subject, I will speak about my own experiment. It was 1966 when Dr. E. Miyamoto and I (11) isolated N-α-acetyl-aspartyl-glutamic acid from bovine brain and determined its structure. We tried to measure the amount of this peptide in the brain by determining glutamic acid after acid hydrolysis, since we had confirmed that glutamic acid was released only from acetylaspartyl-glutamic acid in the acidic fraction prepared from adult rat brain.

We found that there was an unknown compound in immature rat brain which released glutamic acid upon acid hydrolysis. Since the purification was not successful, it was abandoned for about 10 years. Dr. M. Miyake and I (8) then again attacked purification of the compound from about 300 brains of newborn rats. After many trials we succeeded in isolating it as crystal. Elementary analysis, mass spectrography and chemical reactions suggested that the compound was citrylglutamic acid. α- and β-Citrylglutamic acid were synthetized and the structure was finally determined to be β-citrylglutamic acid (Fig. 1). Its concentration is high in neonatal rat brains and in fetal guinea pig brain (8) as shown in Fig. 2 (9). Concentration is also high in adult fish, amphibian and reptile brain and is low in birds and mammalian brains (Table I). We do not know how to interpret our data. There are several possible explanation: the substance may be related to neuronal survival, axon and dendrite formation or to the change from an anaerobic to an aerobic metabolism. Another interesting finding

TABLE I

Concentrations of N-acetylaspartate, N-acetyl-γ-aspartylglutamate and β-Citrylglutamate in the Brains of Various Species

Species	Concentration (μmol/g tissue)		
	NA-Asp	NA-Asp-Glu	β-CG
Fish			
Carp (4)	3.74 ± 0.64	0.03 ± 0.01	0.42 ± 0.06
Amphibian			
Frog[a] (5)	0.10 ± 0.01	1.04 ± 0.05	1.28 ± 0.09
Frog[b] (4)	0.17 ± 0.05	1.11 ± 0.09	1.22 ± 0.01
Reptile			
Turtle (6)	0.88 ± 0.07	0.98 ± 0.23	0.41 ± 0.13
Bird			
Dove (3)	6.99 ± 0.79	0.45 ± 0.06	≤ 0.01
Fowl (4)	7.37 ± 0.12	0.67 ± 0.09	0.02 ± 0.01
Chick (4)	5.25 ± 0.37	0.57 ± 0.03	0.15 ± 0.02
Mammal			
Rat (5)	7.99 ± 0.68	0.74 ± 0.08	0.05 ± 0.01
Guinea pig (4)	7.18 ± 0.49	1.97 ± 0.10	0.02 ± 0.01
Ratbbi (4)	6.60 ± 0.62	0.88 ± 0.11	0.05 ± 0.02

Frog brains were pooled from two animals. Values were corrected on the basis of the recovery rate and expressed as means$\pm$S.D. for the number of determinations given in brackets. Frogs were; [a]*Rana catesbeiana*; [b]*Rana nigromaculata* Hallowell.

concerning β-citrylglutamic acid was obtained from the study of the testis (10). Concentration increased rapidly during maturation of the testis, probably in parallel with sperm formation. When sperm formation was stopped by ligation of the ductus efferentes or transferring the testis back into abdominal cavity, β-citrylglutamic acid disappeared. This fact indicates that this acid is related to the differentiation of sperm cells. As mentioned above, the acid is present in a high concentration, 5 to 10 mmol, in immature brain and in testis. I would welcome suggestions as to what experiments we should plan to clarify the function of this acid.

REFERENCES

1 Barbeau, A. (1961). Proc. VII Int. Cong. Neurol. 2, 135.
2 Bertler, A. and Rosengren, E. (1959). Acta Physiol. Scand. 47, 359.
3 Carlsson, A., Lindquist, M., Magnusson, T., and Waldeck, B. (1958). Science 126, 471.
4 Ehringer, H. and Hornykiewicz, O. (1960). Klin. Wochenschr. 38, 1236.
5 Horiguchi, J., Inami, Y., and Kakimoto, Y. (1985). Jpn. J. Clin. Psychiat. 14, 1091 (in Japanese).
6 Laborit, H., Huguenard, P., and Alluaume, R. (1952). Press. Méd. 60, 206.
7 Delay, J.P. and Deniker, P. (1952). Congrès des Médecins Aliensters et Nourologistes de France, Luxemberg, 497.
8 Miyake, M., Kakimoto, Y., and Sorimachi, M. (1978). Biochim. Biophys. Acta 544, 656.
9 Miyake, M. and Kakimoto, Y. (1981). J. Neurochem. 37, 1064.
10 Miyake, M., Kume, S., and Kakimoto, Y. (1982). Biochim. Biophys. Acta 719, 495.
11 Miyamoto, E., Kakimoto, Y., and Sano, I. (1966). J. Neurochem. 13, 999.
12 Parkinson, J. (1817). An Essay on the Shaking Palsy. London: Sherwood, Neely and Jones.
13 Sano, I., Gamo, T., Kakimoto, Y., Taniguchi, K., Takesada, M., and Nishinuma, K. (1959). Biochim. Biophys. Acta 32, 586.
14 Sano, I. (1959). Adv. Neurol. Sci. 5, 42 (in Japanese).

ARE THERE "GNOSTIC UNITS" IN THE ASSOCIATION CORTEX?

HIDEO SAKATA*

Department of Neurophysiology, Tokyo Metropolitan Institute of Neurosciences, Fuchu, Tokyo 183, Japan

In the early 1960s Hubel and Wiesel (5) found simple cells and complex cells in the striate cortex that were sensitive to the orientation of visual contour. They suggested that these neurons were "building blocks" of visual perception, and that the visual system may be organized in a hierarchical fashion. Konorski (8) proposed the concept of *gnostic units*, by extrapolating the findings of Hubel and Wiesel. He stated "We can assume that perceptions experienced in humans' and animals' lives, are represented not by the assemblies of units (4) but by single units in the highest levels. . . .". He called these levels '*gnostic areas*' and the units responsible for particular perceptions *gnostic units*. Gnostic areas may roughly correspond to the association areas and non-primary sensory areas.

At that time there was no direct evidence that perceptions are really represented by the units of the cerebal cortex. Since the 1970s, single unit analysis of the higher order sensory areas and the association cortex has been done in many laboratories, in alert as well as lightly anesthetized animals. As a result we have now ample evidence that the activity of single units of these areas may be closely related to a particular category of perceptions.

For example, Zeki (23) found the color-coded cells of V4 of the prestriate cortex responded to a particular color in a Mondrian pattern illuminated with three monochromatic lights, but not with the light of that particular color alone, just like we perceive vivid color only with the combination of three monochromatic lights (9). Moreover, Bruce

---

* Present address: 1st Department of Physiology, Nihon University School of Medicine, Ooyaguchi, Itabashi-ku, Tokyo 173, Japan.

et al. (*2*) and Perrett *et al.* (*13*) found that neurons in the bank of the superior temporal sulcus responded to a face. In this review is presented evidence that a class of neurons in the parietal association cortex may represent rotary movement in 3-d space, and the possible relationship of these neurons to the neurons in the superior temporal sulcus responding to the view of body movements (*14*) is discussed.

I. TWO CORTICAL VISUAL SYSTEMS

In classical neuroanatomy, cortical visual areas were confined to the striate and prestriate cortices (areas 17, 18 and 19, or OC, OB and OA). Recent anatomical study by 2-deoxyglucose autoradiography demonstrated that visually active areas are more widespread in the cerebral cortex, which may be subdivided into two major cortical pathways as shown in Fig. 1 (*11*).

One goes ventrally to the inferotemporal cortex (area TE) and then to the lower part of the dorsolateral prefrontal cortex. The other goes dorsally to the posterior part of the inferior parietal lobule (area PG) and then to the frontal eye field and adjacent prefrontal cortex. Previous investigations of neuropsychological ablation experiments suggested

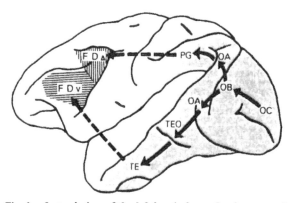

Fig. 1.   Lateral view of the left hemisphere of a rhesus monkey. The shaded area defines the cortical visual tissue in the occipital, temporal, and parietal lobes. Arrows schematize two cortical visual pathways, each beginning in the primary visual cortex (area OC), diverging within the prestriate cortex (areas OB and OA), and then coursing either ventrally into the inferior temporal cortex (areas TEO and TE) or dorsally into the inferior parietal cortex (area PG). From Mishkin *et al.* (*11*).

that these two pathways had different functions. Ablation of the inferotemporal cortex of the monkey caused severe impairment of the task of discriminating visual objects by form (*10*). Ablation of the parietooccipital cortex caused a deficiency in the discrimination of relative distance of a target from a landmark (*16*). On the basis of these observations, Mishkin *et al.* (*11*) proposed a hypothesis that the occipito-temporal pathway is specialized for object perception (identifying *what* an object is), whereas the occipito-parietal pathway is specialized for spatial perception (locating *where* an object is).

One of the major sources of visual input from the prestriate cortex to the posterior parietal association cortex (area 7a or PG) is area MT which is specialized for processing visual movement (*22*) and it should be noted that space vision implicates movement as well as position in space.

II. SINGLE UNIT ANALYSIS IN AREA 7a

In area 7a of alert behaving monkeys, various types of visually responding or eye movement-related neurons were found together with neurons related to manual reaching or hand manipulation (*6, 12*). In a series of single unit analysis of area 7a (*17–20*), we trained the monkeys to fixate their gaze on a stationary spot of light or track a moving spot, giving a drop of juice as a reward. Single unit activity was recorded with a glass-coated platinum-iridium microelectrode inserted into the cortex through the dura mater. A micromanipulator was mounted in a stainless steel chamber implanted in the skull around a trephine hole over the posterior part of the inferior parietal lobule. Eye movement was recorded with a pair of Ag-AgCl pellet electrodes implanted in the rim of orbital bone.

Visual stimuli were given during fixation or tracking. We used a tangent screen to give various visual stimulations such as a linear movement of a slit, changing size of the slit or square and rotation of the slit or square. For depth movement we used a spot of light emitting diode (*LED*) or an acrylic rod illuminated with LED from the inside, which was mounted on a small bench and was moved along a straight rail using a servomotor. The same set of LED spot on the rail was used for the task of visual tracking and fixation in various positions. We also used a small turntable driven by a servomotor to rotate the acrylic rod, LED spots and other stimuli in depth. The rail and turntable were mounted on a stand with a ball-joint to change their orientation in space. A PDP-11/34 computer was used for on-line experimental control and data collection and for off-line data analysis.

Figure 2 shows the site of recording of various types of units of area 7a plotted on a diagram of the cortical surface (19). Figure 2 shows the sites of penetrations of visual fixation (VF) and visual tracking (VT) neurons in area 7a. These neurons are related to eye movement and may be concerned with the perception of position in egocentric space (17), and movement of a visual target in space (18), respectively.

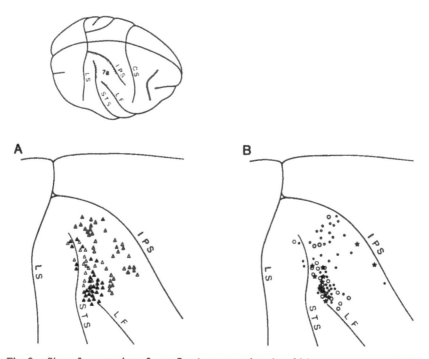

Fig. 2. Sites of penetration of area 7a. A: penetrations in which eye-movement-related neurons were recorded. Open triangles represent those of VF neurons, and solid triangles those of VT neurons. B: penetrations in which passive visual (PV) neurons were recorded. Open circles represent those of depth-movement-sensitive neurons, stars those of rotation-sensitive neurons, and small solid circles those of ordinary PV neurons sensitive to linear movement in the frontoparallel plane. IPS, STS, LS, and LF are the intraparietal, superior temporal and lunate sulcus, and the lateral fissure, respectively. From Sakata et al. (19).

Figure 2B shows the site of penetrations of passive visual (PV) neurons, in which those neurons which responded to depth movement or rotary movement were plotted separately from the ordinary PV neurons which responded to a linear movement of the visual stimulus in the fronto-parallel plane. Most PV neurons were located in the posterolateral part of area 7a, in contrast to those neurons which were related to visual reaching or hand manipulation recorded mainly in the anteromedial part of area 7a, on the posterior bank of the intraparietal sulcus.

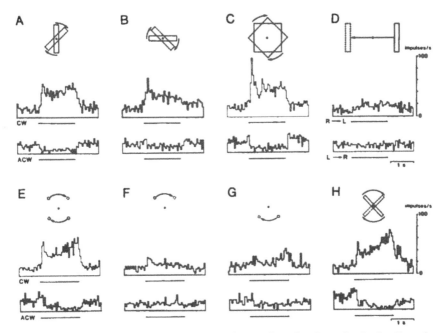

Fig. 3. Responses of an area 7a neuron to the rotation of various visual stimuli on the tangential screen. A: response of the neuron to the rotation of a vertical slit (35 degrees × 7 degrees) at a velocity of 25 degrees/sec. Upper and lower histograms indicate average discharge rate of 10 trials of clockwise and anticlockwise rotations, respectively. B: response to the rotation of the horizontal slit. C: response to the rotation of a square (35 degrees × 35 degrees) at the same speed. D: response to the linear movement of the slit along the horizontal axis at the velocity of 35 degrees/sec toward the right and the left side, respectively. E: response to the rotation of a pair of spots of LEDs 35 degrees apart around the fixation point at the velocity of 50 degrees/sec. F and G: response to the rotation of a single spot in the upper and lower visual fields, respectively. H: response to the rotation of a luminous bar (20 degrees × 1 degree) at the same velocity. From Sakata et al. (20).

III. CHARACTERISTICS OF ROTATION SENSITIVE (RS) NEURONS

Several years ago we happened to find a class of parietal visual neurons which were specifically sensitive to rotary movement of a visual object in space. We studied their functional properties in order to know the neural mechanism of perception of visual rotation. Figure 3 illustrates typical responses of a neuron sensitive to the rotation of various visual stimuli in the frontoparallel plane (20). The cell responded vigorously to clockwise rotation of a vertical slit around the fixation point and its spontaneous discharge was suppressed by anticlockwise rotation of the same slit (A). Its response to the rotation of a horizontal slit was the same in direction but less vigorous than that to the rotation of a vertical slit (B). The cell responded also to rotation of a square in the same direction and this response was a little stronger than the former two (C). In contrast, its response to linear movement of the slit along a horizontal axis was very slight, with the preferred direction toward the left side (D). Vertical movement of the slit was totally ineffective in any part of the visual field. The receptive field of this neuron to rotation was very large, and the response to rotation was independent of position when we shifted the center of rotation 20 degrees from the fixation point in various directions. This means the response to rotation is not due to the local difference in directionality.

In order to differentiate between the effect of changing direction of movement of the visual stimulus and that of changing orientation of the contours, we used a pair or a single spot of light moving around the fixation point. Figure 3 E–G illustrates the response of the same neuron to such stimuli. Responses to the rotation of a pair of spots were almost the same as that to the rotation of an illuminated bar (H), although the response of this neuron to rotation of a single spot was much less than that to a rotating pair of spots (F and G), especially in the upper visual field (F).

IV. RESPONSE TO DEPTH ROTATION

We also found a number of neurons that were sensitive to rotation in depth (20). Figure 4 illustrates the response of a rotation-sensitive

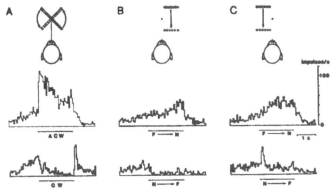

Fig. 4. Responses of an area 7a neuron to the rotation of a visual stimulus in the horizontal plane. A: response of the neuron to the rotation of an acrylic bar (35 cm in length) at the velocity of 45 degrees/sec around the fixation point at the distance of 57 cm. Upper and lower histograms indicate average discharge rate of 10 trials of anticlockwise and clockwise rotations, respectively. B: response of the same neuron to the linear movement of an acrylic bar (12 cm in length) in the depth on the right side of the visual field at the velocity of 10 cm/sec. Upper histogram for the approaching and lower histogram for the receding movement. C: response to the same movement as B on the left side of the visual field. From Sakata *et al.* (*19*).

neuron to the rotary movement of a stimulus in the horizontal plane together with its response to the linear movement in the same plane for comparison. The cell responded vigorously to anticlockwise rotation of an acrylic bar in the horizontal plane, and its spontaneous discharge was strongly suppressed by clockwise rotation (Fig. 4A). It also responded to the approaching movement of a bar on either side of the fixation point (Fig. 4B and C), but its response was much less than that to horizontal rotation. Moreover, the preferred direction of the linear movement in depth was opposite to that of the horizontal rotation on the right side where the movement of anticlockwise rotation was away from the eyes. Thus it is clear that the response of this neuron to rotation in depth was not the sum of the responses to linear movement of the different parts of the rotating stimulus. We found several neurons that were selectively sensitive to horizontal rotation and several others sensitive to rotation in a sagittal plane. In addition, several neurons were found to be sensitive to rotation in more than two planes and responded maximally to rotation in the diagonal planes in between. We compared the responses of several rotation-sensitive neurons to rotation in depth in a binocular viewing condition with those in monocular viewing. In general, the responses in the monocular viewing were less than those in binocular viewing, although in the former they remained sensitive to rotation.

V. RESPONSE TO AMES' WINDOW

There is a famous illusion for depth rotation called "Ames' window" (*1*). It is a trapezoid which appears to be a rectangular window viewed at a slant. When rotated about its vertical axis and viewed with one eye from some distance (several meters) it appears to stop and reverse its direction every 180 degrees. One of the most plausible explanations is that the longer edge always appears to be nearer than the shorter edge in monocular viewing, so that the longer edge looks as if it moves in front of the axis of rotation when it actually moves behind the axis.

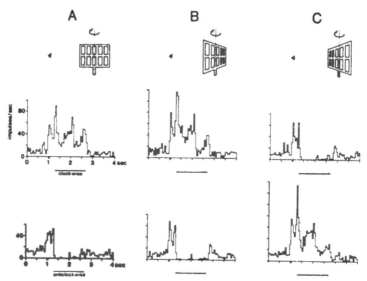

Fig. 5. Reversal of the preferred direction of an RS neuron in response to the trapezoid window. A: response to the rotation (180 degrees) of a symmetrical window for control. Upper histogram is an average response to the clockwise rotation; lower histogram an average response to the anticlockwise rotation. B: response to the rotation of a trapezoid window, 180 degrees, while the longer edge moved in front of the axis. C: response to the rotation of a trapezoid window, while the longer edge moved behind the axis. Note that better response was obtained with the anticlockwise rotation.

Figure 5 illustrates an example of RS neurons which displayed a reversal of preferred direction in response to the rotation of the trapezoid window. The preferred direction of this neuron was clockwise, as illustrated by the response to 180 degree rotation of a symmetrical window at the distance of 150 cm in monocular viewing (Fig. 5, left). Preferred direction was clockwise when the longer edge of the trapezoid window moved in front of the axis of rotation. But the neuron preferred anticlockwise rotation when the longer edge moved behind the axis, showing a reversal of directionality, just as we perceive the reversal of rotation of that window in monocular viewing.

The fact that the response of rotation-sensitive neurons corresponds very well to the subjective experience of illusion of rotation in depth strongly suggests that the activity of these neurons is directly related to the perception of rotation in space.

VI. REPRESENTATION OF MOVEMENT AND SHAPE BY A SINGLE UNIT

What is the relation of these parietal neurons to the perception of real objects? The recent anatomical study of Seltzer and Pandya (21) demonstrated that there is a strong projection from area PG to the bank of STS in the temporal cortex which may be mixed with the projection from the inferotemporal cortex.

This is the area where Bruce et al. (2) as well as Perrett et al. (13) found neurons responding to the face. They also found neurons responding better to the profile than to the full face, as well as neurons preferring several other orientations of the face (3, 15).

Recently Perrett et al. (15) found in the upper bank and fundus of STS many neurons which responded to particular classes of body movements. Figure 6 illustrates a neuron that was sensitive to the rotation of the head. The cell responded more to the rotation of the head from left or right profile to confront the monkey with a full face than to rotation of the head from full face away to left or right profile. This neuron did not respond to the rotation of any other 3-D object than the head. Therefore, these neurons seemed to have conjoint sensitivity to form and motion. And it is likely that they receive information about movement from RS neurons of the parietal cortex. The authors also

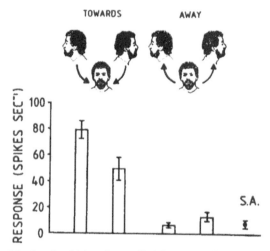

Fig. 6. Sensitivity of one cell of the temporal cortex to head rotation towards the monkey. The mean and standard error (spikes/sec) of responses to rotation of the head between face and profile are shown for cell A043. The cell responded more to the rotation of the head from left or right profile to confront the monkey with a full face than to rotation from full face away to left or right profile ($p<0.01$ each comparison). Responses to rotations from face to profile were not different from spontaneous activity (S.A.). Number of trials per condition=8, 5, 4, 5, 7. Overall effect of conditions, $F=29.2$, $df=4.24$ $p<0.0001$. From Perrett et al. (14).

recorded a group of temporal neurons responsive to body movement which preferred the view of a person walking in a particular direction.

The body is a multiply articulated object and the movement of a walking person is the combination of multiple rotations and a linear translative movement. The pattern of a walking person may be represented by a limited number of spots as demonstrated by Johansson (7). Several types of RS neurons and a class of visual tracking neurons may be activated by such a pattern of visual stimulus.

Therefore, it is plausible that these different classes of parietal neurons converge upon a single temporal neuron responding to the view of a walking person. A single unit in the temporal association cortex may thus represent a dynamic scene as a whole.

REFERENCES

1 Ames, A. (1951). Psychol. Monogr. 65 (7, Whole No. 324).
2 Bruce, C., Desimone, R., and Gross, C.G. (1981). J. Neurophysiol. 46, 369.

3 Desimone, R., Albright, T.D., Gross, C.G., and Bruce, C. (1984). *J. Neurosci.* **4**, 2051.

4 Hebb, D.C. (1957). *The Organization of Behavior: A Neuropsychological Theory.* New York: Wiley.

5 Hubel, D.H. and Wiesel, T.N. (1962). *J. Physiol.* **160**, 106.

6 Hyvarinen, J. and Poranen, A. (1974). *Brain,* **97**, 673.

7 Johansson, G. (1975). *Sci. Am.,* June 76.

8 Konorski, J. (1967). *Integrative Activity of the Brain, An Interdisciplinary Approach.* Chicago: The University Chicago Press.

9 Land, E. and McCann, J. (1971). *J. Opt. Soc. Am.* **61**, 1.

10 Mishkin, M. (1972). Cortical visual areas and their interactions. In *Brain and Human Behavior* ed. Karczmar, A.G. and Eccles, J.C., p. 187. Berlin: Springer-Verlag.

11 Mishkin, M., Ungerleider, L.G., and Macko, K.A. (1983). *Trends. Neurosci.* **5**, 414.

12 Mountcastle, V.B., Lynch, J.C., Georgopoulos, A., Sakata, H., and Acuna, C. (1975). *J. Neurophysiol.* **38**, 871.

13 Perrett, D.I., Rolls, E.T., and Caan, W. (1982). *Exp. Brain Res.* **47**, 329.

14 Perrett, D.I., Smith, P.A.J., Mistlin, A.J., Chitty, A.J., Head, A.S., Potter, D.D., Broennimann, R., Milner, A.D., and Jeeves, M.A. (1985). *Behav. Brain Res.* **16**, 153.

15 Perrett, D.I., Smith, P.A.J., Potter, D.D., Mistlin, A.J., Head, A.S., Milner, A.D., and Jeeves, M.A. (1985). *Proc. Roy. Soc. Lond.* **B 223**, 293.

16 Pohl, W. (1973). *J. Comp. Physiol. Psychol.* **82**, 227.

17 Sakata, H., Shibutani, H., and Kawano, K. (1980). *J. Neurophysiol.* **43**, 1654.

18 Sakata, H., Shibutani, H., and Kawano, K. (1983). *J. Neurophysiol.* **49**, 1364.

19 Sakata, H., Shibutani, H., Kawano, K., and Harrington, T.L. (1985). *Vision Res.* **25**, 453.

20 Sakata, H., Shibutani, H., Ito, Y., and Tsurugai, K. (1986). *Exp. Brain Res.* **61**, 658.

21 Seltzer, B. and Pandya, D.N. (1984). *Exp. Brain Res.* **55**, 301.

22 Ungerleider, L.G. and Desimone, R. (1986). *J. Comp. Neurol.* **248**, 190.

23 Zeki, S. (1980). *Nature* **284**, 412.

ISOLATED PREPARATIONS OF NEWBORN RAT CENTRAL NERVOUS SYSTEM

MASANORI OTSUKA

Department of Pharmacology, Faculty of Medicine, Tokyo Medical and Dental University, Tokyo 113, Japan

Isolated preparations of the mammalian central nervous system (CNS), such as brain slices and dissociated cultured neurons, provide a useful means for physiological and pharmacological studies (*12*). Here I describe several different preparations of the CNS isolated from newborn rats that we are using to study the functions of neurotransmitters and the effects of drugs.

I. ISOLATED SPINAL CORD

About 10 years ago we developed an isolated spinal cord preparation of the newborn rat (*7*). When the preparation is perfused by artificial cerebrospinal fluid (CSF) at 27°C, it can be kept alive for more than 10 hr. The reasons why the preparation stays alive for a relatively long time are probably that the diameter of the spinal cord is small (about 2 mm or less), which allows an adequate oxygen supply by diffusion from the perfusion solution, and that the oxygen consumption of neonatal CNS tissues is less than that of adult tissues (*2*). With this preparation, it is easy to record the monosynaptic and polysynaptic reflexes from the ventral root or intracellular potentials from the motoneurons (*3*), and to observe the effects of ion concentrations and of drugs.

II. BRAINSTEM-SPINAL CORD PREPARATION

Recently, by a modification of the isolated spinal cord preparation, an isolated brainstem-spinal cord preparation of the newborn rat was developed (*6, 10, 11*). It is possible in this preparation to record the spontaneous rhythmic respiratory activity either from the cervical ventral root or phrenic nerve and to observe the effects of drugs (enkephalins, substance P, *etc.*) and physicochemical environment (pH, temperature, ion concentration) (*6, 10*). It is also possible to observe descending inhibition of a certain type of spinal reflex (*1*). When lungs and trachea are kept connected with this preparation through a vagal nerve, the inflation of the lungs by gas pressure reduces the respiratory frequency (*5*). This inhibitory respiratory reflex resembles Hering-Breuer's inflation reflex. The reflex was partly blocked by bicuculline or strychnine, which suggests that γ-aminobutyric acid (GABA) and glycine are involved in the reflex as neurotransmitters.

III. STUDY OF PAIN *IN VITRO*

Studies of pain and analgesic drugs have often been made with intact animals. Such experiments have many disadvantages: 1) the experimental animals suffer from pain, which causes an ethical problem; 2) it is not possible to control the drug concentration at the site of action in the CNS; 3) quantitative analysis of nociceptive responses (*e.g.*, vocalization, writhing response) is sometimes difficult; 4) in testing of the drug effects it is often difficult to differentiate the analgesic effect from motor paralysis; 5) the drugs must pass the blood-brain barrier to act on the CNS. To overcome these disadvantages, we are using isolated spinal cord-tail preparation of the newborn rat, in which the tail is attached to the cord through the filum terminale and the spinal nerves below the S3 segment (*13, 14*) (Fig. 1). When the tail is pinched by forceps, the movement of which is controlled by a plastic syringe, gas pressure, a solenoid valve, and an electronic stimulator, this stimulus causes a depolarizing response of the lumbar ventral root, referred to as the tail-pinch potential (*13*). The tail-pinch potential is depressed by morphine and enkephalins, and the effects of these drugs are reversed by an opiate antagonist, naloxone (Fig. 2). By contrast, the monosynaptic reflex is little affected by these drugs. It is also possible to stimulate the tail with chemicals. For this purpose, the superficial layer of the skin of the tail is removed and the tail is perfused with artificial CSF (*14*). Application to the tail of capsaicin, a pungent substance that selectively stimulates sensory C-fibers, induces a depo-

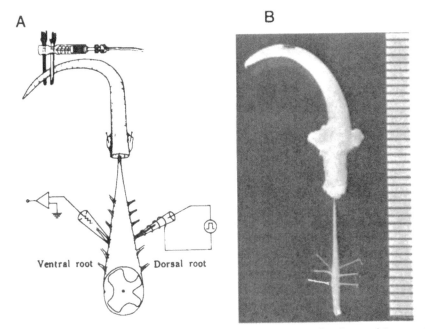

Fig. 1. Isolated spinal cord-tail preparation of the newborn rat. A: scheme of the experimental arrangements. The spinal cord below the thoracic part (Th12) was isolated together with the tail, which remained connected to the cord through the filum terminale and coccygeal nerves. A ventral root (L3-L5) was sucked into the recording suction electrode, which tightly fitted the ventral root. The dorsal root of the corresponding segment was placed in the stimulating suction electrode. The tail at a point 3–4 mm distant from the tip was covered with nylon mesh and placed between forceps. The movement of the forceps was controlled by a syringe connected to a nitrogen tank through a solenoid valve. When the solenoid valve was opened for the desired period (0.02–0.8 sec) by a pulse generator and pressure (0.2–1.6 kg/cm^2) was applied to the piston, the glass bar attached to the piston pushed one end of the forceps; when the solenoid valve was closed, the gas pressure was released and the glass bar was returned to the original position by a spring inside the syringe. Thus, a pressure stimulus of controlled strength and duration could be applied to the tail. B: photograph of the preparation. The place for pressure stimulation is marked in color. One division in the scale is 1 mm. From ref. *13*.

larizing response of the lumbar ventral root. This capsaicin-induced nociceptive response is depressed by morphine, Met-enkephalin, and many other drugs (*15*).

E-type prostaglandins are known to sensitize primary afferent terminals to noxious stimuli in peripheral tissues. As could be predicted, the pretreatment of the tail with prostaglandin E_1 or E_2 (4 μM) by

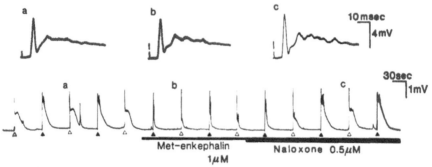

Fig. 2. Effects of [Met⁵]enkephalin and naloxone on the tail-pinch potential and spinal reflexes. Single shock stimulation was given to the dorsal root (L4) every 2 min at the time indicated by △ under the lower tracing and the pressure stimulus (1 kg/cm², 0.4 sec applied to the syringe) was given to the tail alternately at the times indicated by ▲. The potential was recorded from the ipsilateral L4 ventral root and displayed on a pen-recorder (lower trace). Upper tracings show the oscilloscope records of monosynaptic and polysynaptic reflexes taken at the times indicated on the lower tracings. [Met⁵]enkephalin was applied during the period indicated by a black horizontal bar under the lower tracing. During the period indicated by the thick horizontal bar, naloxone was further added in the continual presence of [Met⁵]enkephalin. From ref. 13.

perfusion greatly potentiates the capsaicin-induced nociceptive reflex (16).

IV. SUBSTANCE P AS A PAIN TRANSMITTER

Substance P is found in certain subpopulations of primary afferent C-fibers. It is generally believed that C-fibers are important in conveyance of the sensation of slow pain to central neurons. As to the roles of SP in primary afferent C-neurons, there is evidence that SP is released from the central terminals of primary afferent C-fibers and has an excitatory effect on spinal neurons (8). On the other hand, in the inferior mesenteric ganglion of the guinea pig, SP is released from the peripheral terminals of axon collaterals of certain visceral primary afferents and produces a slow excitatory postsynaptic potential (EPSP) in nerve cells (4). Based on these results and others, a hypothesis emerges that SP that is released from certain sensory C-fiber terminals in the dorsal horn serves as a transmitter producing slow EPSPs in the dorsal horn cells to transmit slow pain signals. To test this hy-

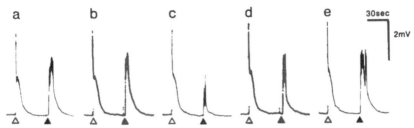

Fig. 3. Effect of [D-Arg^1, D-Pro^2, D-Trp^{7,9}, Leu^{11}]SP on the capsaicin-induced nociceptive reflex in an isolated spinal cord-tail preparation of a newborn rat. $\triangle$ spinal reflexes induced by a single shock stimulus (0.1 msec, 27V) given to the ipsilateral L4 dorsal root; $\blacktriangle$ capsaicin-induced potential evoked by application of the drug (0.5 μM, 0.4 sec) to the tail. The potential was recorded extracellularly from the L4 ventral root and displayed on a pen-recorder. a and b: two successive control records in normal medium immediately before the SP antagonist was added. c: 5 min after the SP antagonist (10 μM) was added to the solution perfusing the spinal cord. d and e: 43 and 58 min after removal of the antagonist. From ref. *14*.

pothesis, the effects of SP antagonists on two kinds of C-fiber reflexes were examined.

In the first type of experiment, the tail was stimulated with capsaicin in an isolated spinal cord-tail preparation of newborn rats, and the nociceptive reflex was recorded from the lumbar ventral root. The capsaicin-induced nociceptive reflex was markedly and reversibly depressed by two SP antagonists, [D-Arg^1, D-Pro^2, D-Trp^{7,9}, Leu^{11}]SP (10 μM) (*14*) (Fig. 3) and [D-Arg^1, D-Trp^{7,9}, Leu^{11}]SP (Spantide) (16 μM) (Otsuka and Yanagisawa, in preparation). In the second type of experiment, a cutaneous (saphenous) nerve was stimulated with single or double shocks of high intensity, which induced a slowly conducting afferent volley in the L3 dorsal root and a slow depolarizing reflex in the L3 ventral root. This slow depolarizing response induced by intense stimulation of the saphenous nerve was inhibited by Spantide (16 μM) (Nussbaumer *et al.*, in preparation).

V. INHIBITION OF NOCICEPTIVE INPUT

There are inhibitory mechanisms in the substantia gelatinosa of the spinal cord by which pain signals can be controlled. GABA and enkephalins, both of which occur in high concentrations in the superficial layers of the dorsal horn, may be the neurotransmitters involved. In support of this notion, the tail-pinch potential is potentiated by naloxone (0.5 μM), an opiate antagonist, and bicuculline (1 μM), a GABA antagonist (9, 13).

VI. CONCLUSIONS

The isolated preparations of the CNS described here are relatively easy to prepare and economical, and therefore may open up many further possibilities. For example, organ culture of these preparations might enable us to study long-term changes in the CNS. Also, autonomic reflexes involving the visceral organs might be studied in vitro. Furthermore, the higher centers of the newborn rat could also be studied in vitro with similar techniques if the ethical problems are solved.

REFERENCES

1 García-Arrarás, J.E., Murakoshi, T., Yanagisawa, M., and Otsuka, M. (1986). Brain Res. 379, 188.
2 Itoh, T. and Quastel, J.H. (1970). Annual Report of the Center for Adult Diseases. vol. 10, p. 75. Osaka: Center for Adult Diseases.
3 Konishi, S. (1982). In Advances in Pharmacology and Therapeutics II, ed. Yoshida, H., Hagihara, Y., and Ebashi, S., vol. 2, p. 255. Oxford: Pergamon Press.
4 Konishi, S. and Otsuka, M. (1985). J. Physiol. 361, 115.
5 Murakoshi, T. and Otsuka, M. (1985). Neurosci. Lett. 62, 63.
6 Murakoshi, T., Suzue, T., and Tamai, S. (1985). Br. J. Pharmacol. 86, 95.
7 Otsuka, M. and Konishi, S. (1974). Nature 252, 733.
8 Otsuka, M. and Konishi, S. (1983). Trends Neurosci. 6, 317.
9 Otsuka, M. and Yanagisawa, M. (1987). Acta Physiol. Hung. 69, 363.
10 Suzue, T. (1984). J. Physiol. 354, 173.
11 Suzue, T., Murakoshi, T., and Tamai, S. (1983). Biomed. Res. 4, 611.
12 Yamamoto, C. and McIlwain, H. (1966). J. Neurochem. 13, 1333.
13 Yanagisawa, M., Murakoshi, T., Tamai, S., and Otsuka, M. (1984). Eur. J. Pharmacol. 106, 231.
14 Yanagisawa, M. and Otsuka, M. (1984). Proc. Jpn. Acad. 60B, 427.
15 Yanagisawa, M. and Otsuka, M. (1985) Jpn. J. Pharmacol. 39 (Suppl.), 73P.
16 Yanagisawa, M., Otsuka, M., and García-Arrarás, J.E. (1986). Neurosci. Lett. 68, 351.

NEUROCHEMICAL STRATEGIES TO ELUCIDATE THE MECHANISM OF LEARNING AND MEMORY: COMMENT

YASUZO TSUKADA

Department of Physiology, School of Medicine, Keio University, Tokyo 160, Japan

"To elucidate the mechanism of learning and memory on molecular and cellular levels. . . . . ."

One goal of brain research is to explain the higher functions of the brain in terms of neurobiology. At present, molecular and cellular approaches to elucidate the neural mechanism of learning and memory with the use of experimental animals is an effective strategy. In particular, it is essential for the study of long-term memory to examine the biochemical changes in the brain that accompany lasting changes in learned behavior in animals such as fish, chickens, mice, rats, and monkeys.

It is generally recognized that when protein synthesis in the brains of fish and mice is inhibited experimentally, learning acquisition does not occur; that is, protein synthesis may be necessary for memory consolidation. We have found that the administration of an inhibitor of protein synthesis to the brains of chickens prevents their becoming imprinted. This strongly suggests that for memory to be consolidated as long-term memory, synaptogenesis, which is accompanied by protein synthesis, is needed. It is important to find whether the protein synthesized is necessary for a particular synaptic element on the synaptogenesis, and whether this protein synthesis is regulated on the level of the gene or whether it arises simply from acceleration of protein synthesis in the cytoplasm. It is also important to find what kinds of neurotransmitters are contained in the newly formed synapses.

For research into synaptogenesis, I think that the cellular biological approach with cultured and grafted nervous tissues is also a promising field.

It is possible to think of the phenomenon of long-term potentiation (LTP) that has been studied in the hippocampus and elsewhere as being one basic pattern of memory, but is it also possible to see it as a model of short-term memory instead of long-term memory in which there are changes in the efficiency of transmission because of chemical modifications at the synaptic membrane? So, in LTP, protein synthesis may not be necessary, but there is no doubt that LTP is one aspect of the plasticity of the nervous system.

There is another strategy to elucidate the mechanism of higher nervous functions such as learning and memory: to use experimental models of abnormal development of the brain (of rats and monkeys) such as cretinism and phenylketonuria, which are good examples of mental retardation. Results so far of analysis of experimental model animals show that there is insufficient myelin formation in the brain and particularly in the cerebrum accompanying poor learning ability tested by discriminative learning tasks. It is also possible that synapse formation followed by learning is suppressed because of decreased protein synthesis.

Research on the development and differentiation of the brain on the molecular and cellular levels is an urgent and important theme to achieve a breakthrough in brain research.

A COMMENT ON DEVELOPMENT AND DIFFERENTIATION OF THE BRAIN

HIROSHI YOSHIDA

Department of Pharmacology I, Osaka University School of Medicine, Osaka 530, Japan

One of the most important and interesting problems facing neuroscience in future is to know how the mechanism of neuronal circuits, functional connections between specific neurons, is formed and maintained. The development and differentiation of the nervous system are expressed by cell proliferation, cell death and elimination of neurites morphologically, and are also closely related with plasticity in the physiological field.

The development and differentiation are regulated by environmental factors with genetic control, among which are the environmental agents known as trophic factors. Nerve growth factor is one example of a trophic factor and many other kinds—growth factor, synapse formation factor and so on—will be discovered in future research. It is difficult to speculate on the number of neurotrophic factors present in the brain, but I think it may be more than the number of neurotransmitters. I assume that the many kinds of trophic factors will be divided into those of a specific and non-specific nature in future. Among these, I have a great deal of interest in the specific trophic factors which are formed in specific neurons and which show a selective effect on these specific neurons, such as neurotransmitters.

A deficiency in some specific trophic factors in the brain may cause degeneration on a specific part of the nervous system in the brain as in Alzheimer's disease, though another possibility such as the accumulation of toxic substance(s) may also be considered a cause of degenerative disorders.

Thus, one of the most urgent problems in the search for trophic factors is to establish their assay system. Morphological changes seem to be the most successful means of doing this at present, but this is a troublesome method. Other systems for screening these factors are therefore required. Induction of a microtubular system and inhibitory activity on a certain kind of serine protease may be considered. Further work, however, seems necessary to establish the assay sysem for neurotrophic factors, because the general properties for the factors are not known and there clearly appear to be factors with different properties.

ONE OPINION IN REGARD TO THE PROMOTION OF NEUROSCIENCE

YUTAKA SANO

Department of Anatomy, Kyoto Prefectural University of Medicine, Kyoto 602, Japan

Nerve cells have processes which bifurcate in a complicated pattern, showing morphological characteristics markedly different from those of general somatic cells. The most important goal of the neurosciences on the molecular level is to clarify the mechanism by which complicated neuronal circuits are formed by neurons with highly specific structure to perform a variety of central nervous functions.

Neurons composing the nervous system have greatly varied cytological structures: some of them have long processes of more than 1 m in length, but others have short processes of only about 10 μm. Consequently, investigators must first realize that there is a great difference between a study of an organ composed of aggregated approximately similar cells, *e.g.*, liver composed of hepatocytes, and a study of brain.

Neurological study has a long history, most of it spent by investigators in groping. Neurobiologists recognize that the study of the nervous system started in modern science with the discovery of the silver impregnation method by Camillo Golgi (1873). In those days, neuroscience was established on the basis of morphological studies using the silver impregnation technique.

The silver impregnation method enables us to clearly visualize the perikarya and the two kinds of processes, dendrites and axons. Most neurobiologists believe that the images of black impregnated neurons truly represent the features of nerve cells existing *in situ*. Simplifying the images, a model scheme of a neuron can be produced, but no such neuron with the simple processes seen in the model exists *in vivo*.

Recently methods by which neurons are labeled by the intracellular injection of tracer substances, such as horseradish peroxidase, plant lectin, *etc.*, using a micropipette electrode have been developed. The three-dimensional reconstruction of the axonal arborizations of the labeled neuron can be produced by tracing with a computer graphic system. The image of a neuron yielded by these modern methods is more complicated than that obtained by the silver impregnation method, and is highly varied. These results suggest that the neuronal image obtained so far by the silver impregnation is incomplete, revealing a great disparity between the image of the impregnated neuron and that of the real neuron.

In neuroscience, as in all natural scientific studies, investigators have given priority to detecting the neurobiological laws on the basis of the minute analysis of common mechanisms prevailing in various neuronal activities. Today, we have many scientific data on various neuronal phenomena, *e.g.*, conduction of impulses, neurotransmission, axonal flow, plasticity, *etc.*. In neuroscience, however, inspection of commonalities alone is overly simplistic. Until the specific identity of each neuron among the greatly varied forms has been determined and the neuronal images established so far by model schemes are thrown out, we cannot approach a real understanding of the brain.

STATE- AND TISSUE-DEPENDENCE OF NEURONAL FUNCTION

MOTOY KUNO

Department of Physiology, Kyoto University, Faculty of Medicine, Kyoto 606, Japan

The early approach toward neuroscience at a cellular level was initiated by the analysis of ionic mechanisms underlying excitation. This was followed by detailed investigations on every step involved in synaptic transmission. These studies provided our current concept as to how excitation occurs in a given neuron and how an impulse is transmitted from one neuron to another. Furthermore, through the analysis of a neuronal network, we can now predict which neuron groups will be synaptically excited or inhibited by a particular sensory stimulus in a stereotypical fashion.

While the "stereotype" is one feature of the neuronal function which executes precise signaling, the state of neuronal function can also be modifiable, depending upon the past experience. Thus, a neuronal response to a stimulus may eventually disappear when the same stimulus is repeatedly applied (habituation). Also, a stimulus that was initially ineffective may become effective in evoking a response, if applied repeatedly in conjunction with another type of stimulus (classical conditioning). Therefore, neuronal function is state-dependent. State-dependence requires the storage of neuronal information by which a novel neuronal function is expressed. How state-dependent alterations occur in neuronal function is one of the tasks prospectively assessed by neuroscience.

Another puzzling behavior of the nervous system is specificity. There are a large number of examples of neuronal specificity. Specific neuronal connections are determined in a preprogrammed fashion at embryonic stages. Also, a particular group of neurons migrates from its initial site to a specific region during early developmental stages.

The final destination of neurons is now known to depend upon their initial site. Thus, neuronal specificity can be dictated by information of the tissue to which the neurons are first exposed. In other examples, specificity appears to be determined by the final destination tissue. For example, nicotinic acetylcholine receptors are present in skeletal muscle but not in smooth muscle. Similarly, sodium channel molecules are present in excitable tissues but not in non-excitable tissues. Therefore, it is evident that gene expression in the nervous system is tissue-dependent as well as state-dependent. How tissue-dependent specificity is achieved is another task facing neuroscience in the future.

It is conceivable that the development of any novel state of neuronal function or the acquisition of any neuronal specificity is always associated with the synthesis of specific mRNAs. One can thus speculate that the synthesis of new mRNAs may occur during plastic changes of neuronal function, depending upon the past experience. Also, the preprogram of neuronal development at embryonic stages may be determined by the tissue-specific synthesis of mRNAs. Regulation of gene expression such as alternative RNA splicing events would provide great insight into the molecular basis of neuronal specificity and plastic changes. The genetic approach is certainly a powerful strategy for neuroscience in the future.

NEUROSCIENCE IN THE STUDY ON BRAIN AGING AND DEMENTIA

MASANORI TOMONAGA

Department of Neuropathology, Institute of Brain Research, Faculty of Medicine, University of Tokyo, Tokyo 113, Japan

Thank you very much for inviting me to this symposium. I am very much stimulated. I am now working on neuropathological problems, particularly brain aging and dementia, and I think this area is the final target of neuroscience. Recent technological developments in neuroscience are of help in this research; for example, nerve growth factor (NGF), which is rich in the septohippocampal cholinergic system of the brain and rich in astroglial cells in the brain may open the door to treatment and eventual prevention of dementia and other degenerative diseases. Dr. Shooter showed that NGF regulates the production of microtubules and tau protein in the developmental stage of neural construction. It is known now that tau protein, in phosphorylated form, is a major component of paired helical filaments (PHF), the fibrillary protein of Alzheimer's neurofibrillary tangles, which is a hallmark of Alzheimer's disease. Another example is an increase of glial cell population in the aged brain; this finding might be a compensatory phenomenon. The trophic interrelationship between neuron and glia is very important in the aging brain. The approach made by Professor Fujita will open another door. Dr. Diamond in California showed that the brain of Albert Einstein contained several hundred fold glial cells more than the normal (control).

I examined aged elite brains and found very well preserved or much more developed dendritic arborization in the hippocampus than in younger brains. This plastic response is very important for the aged brain to preserve its function in later years. Recently psychological and positron emission tomography (PET) studies of aged people have shown that their intelligence continues to increase throughout their lifetime. This ability is called "crystallized intelligence." I think Sir John Eccles is a typical example of this phenomenon. This mystery will hopefully be elucidated in the near future.

LEARNING ABOUT LEARNING

NOZOMU SAITO

Department of Physiology, Dokkyo University School of Medicine, Mibu, Tochigi 32-102, Japan

Learning is a complex neural process involving acquisition, storage and retrieval of information, to mention just a few of the subprocesses involved. Recent advances in several of these fields are extensive. In contrast, little is known of the physiological basis of the retrieval process. In the field of ethiological observations, there are some relevant models of avian brain that enable our understanding of the mechanism by which information is stored and retrieved.

Songbirds such as the canary and zebra finch memorize the song of conspecific adults as an auditory template during a critical period early in life, and later retrieve the memorized patterns of song and develop them by utilizing auditory feedback. During the early stage of song development, immature patterns of a song are voiced by the vocal organ, the syrinx. An afferent signal of a vocal gesture of the motor of the syrinx feeds back to the control center, perhaps by an internal feedback path. Other external feedback signals, mediated by voice, come to the center and meet there with the internal feedback. The internal and external feedback may intersect somewhere during retrieval of the memory pattern to create an auditory template of the song. Neurons in one of the telencephalic nuclei (HVc) controlling song have recently been shown to respond to acoustic stimuli. The auditory response of some neurons in this nucleus permits great flexibility in manipulating complex feedback signals. The results of this are presented here.

Conditioning stimuli applied to the syringeal motoneuron in the medulla modulate central auditory responses in the HVc (*1*). On the other hand, the syringeal motoneuron responds to complex auditory stimuli such as song. This response has been shown to depend on the HVc and to be very selective to the conspecific song (2). To this extent, vocal perception seems to depend on vocal production, and this occurs in either central or peripheral sites of the avian song control system. Close linkage of production and perception leads to a hypothesis that the conspecific song may be perceived as a set of vocal gestures whose members are discriminated from other environmental sounds.

This hypothesis of the avian brain might contribute a clue to the understanding of some aspects of the process of learning, including perception or recognition that reflects the physiological basis of the retrieval process of information.

REFERENCES

1 Maekawa, M. and Saito, N. (1985). *J. Physiol. Soc. Jpn.* **74**, 500.
2 William, H. and Nottebohm, F. (1985). *Science* **229**, 279.

SLEEP AS AN ADVANCED CEREBRAL FUNCTION

SHOJIRO INOUÉ

Institute for Medical and Dental Engineering, Tokyo Medical and Dental University, Tokyo 101, Japan

The cerebrum, the most highly developed integrative and regulatory organ, seems to be unable to maintain an active state and regularly requires a special state of rest, *i.e.*, sleep. Sleep is not observable in artificial device systems, no matter how sophisticated they are, nor is it seen in plants or lower animals. Hence sleep is closely connected with one of the high-order cerebral functions. However, neuroscientists pay rather little attention to the "resting" brain and sometimes overlook the function of sleep. I should like to emphasize the importance of challenging the mystery of sleep for a deeper understanding of the cerebral functions.

The regulatory mechanism of sleep is largely unknown to date. There are various unsettled issues on the localization of the sleep-regulatory center in the central nervous system. The orthodox neurophysiology of sleep seems to pose many obstacles to further progress. On the other hand, a somewhat heterodox approach to sleep mechanisms has revealed in the past two decades that sleep is humorally regulated by a number of endogenously liberated biochemical factors. Thus the classical concept on the humoral regulation of sleep is revived and firmly established.

A wide variety of candidate substances, which are called "endogenous sleep substances," are known to modify either non-REM sleep or REM sleep or both, if they are administered at a certain time of day with an optimal dosage through an exogenous route. The 8th Taniguchi Symposium on Brain Science was timely as it offered the first opportunity for integrating the knowledge gained in broadly diverse experimental results (*1*).

Since then, rapid progress has been occurring in this new field of sleep research. Discoveries of novel sleep-modulating compounds have been reported one after another. Their non-specific nature has been gradually recognized by researchers and has puzzled them as to what this signifies regarding the existence of a sleep-specific endogenous factor. Meanwhile, attempts have been made to analyze the action of putative sleep substances on neuronal activities of the sleep-regulating areas of the brain. Furthermore, the restitutive function of sleep has begun to be evaluated in consideration of the thermoregulatory and immunoactive property of some sleep substances. Trial clinical applications of synthetic compounds have been made to cure insomniacs with special reference to their circadian sleep-waking rhythms.

Thus, the humoral regulatory aspect of sleep research seems to be a promising and expanding field of neuroscience, which may in future contribute much to an understanding of the mechanism of sleep and eventually to the function of the brain.

REFERENCE

1 Inoué, S. and Borbély, A.A., eds. (1985). *Endogenous Sleep Substances and Sleep Regulation.* Tokyo and Utrecht: Japan Sci. Soc. Press and VNU Science Press BV.

IDENTIFICATION OF SPECIFIC SUBSTANCES IN NEURAL DEVELOPMENT

KUNIHIKO OBATA

Department of Pharmacology, Gunma University School of Medicine, Maebashi 371, Japan

Many substances are probably involved in axonal growth and synapse formation at the developmental stage of the nervous system. Furthermore, such substances may participate in normal function and regeneration of the adult nervous system. Nerve growth factor (NGF) plays various roles in some peripheral and cholinergic central neurons. But corresponding factors have not been found for other neurons. Recently several cell adhesion molecules (CAMs) and extracellular matrix substances are believed to support the directed and fasciculated growth of the axons. However, no substance for cell-cell recognition or position marking as proposed by Sperry's chemoaffinity hypothesis on neural connection has been discovered.

One approach to discovering the unidentified factors is to employ an elementary phenomenon such as cellular adhesion or neurite outgrowth for the *in vitro* assay and to look for active substances in the embryonic material. Another approach is to determine the molecules which are specifically expressed at certain developmental stages in the restricted region and then to explore their physiological function. For detection of such substances, not only biochemical analysis but also the monoclonal antibody technique is useful. For further investigation of their chemical nature and specific gene expression recombinant DNA technology can be applicable.

By systematic analyses of the proteins in the embryonic chick optic tectum using quantitative two-dimensional gel electrophoresis, we have disclosed Drebrin, a family of developmentally regulated brain proteins (Shirao and Obata, *J. Neurochem.* **44**, 1210, 1985 and *Dev. Brain Res.* **29**, 233, 1986). They are three acidic 95–110 kD proteins which are highly homologous on the peptide map and are indistinguishable by five monoclonal antibodies. Each member appears and disappears successively during neural development. A cDNA library was prepared from chick embryo mRNA with an expression vector and fusion proteins were screened with anti-Drebrin antibodies. Several Drebrin-cDNA clones were isolated and their characterization is under way.

In order to survey chemical characteristics of the early embryonic neural tissue, we produced monoclonal antibodies against the membrane fraction of the neural tube isolated from 3-day chick embryos. Each of them specifically binds the neural tissue, the neural crest, non-neural epithelial cells or the like. Characterization of the antigens and their comparison with the CAMs and other substances are in progress.

REGENERATION OF THE CEREBELLOFUGAL PROJECTION IN KITTENS

SABURO KAWAGUCHI

Institute for Brain Research, Faculty of Medicine, Kyoto University, Kyoto 606, Japan

In contrast to the current concept of abortive regeneration of mammalian central axons, the occurrence of marked, functionally active, regeneration of the cerebellofugal projection was proved in kittens after complete transection of the decussation of the brachium conjunctivum (BCX) (Kawaguchi *et al., J. Comp. Neurol.* **245**, 258, 1986).

Complete transection of the BCX was achieved by pushing down an edged U-shaped wire to the brainstem base at the midline, and the wire was left *in situ* to mark the lesion. Later, horseradish peroxidase was injected into the lateral and interpositus nuclei to label anterogradely the cerebellofugal projection arising from these nuclei up to the terminals. In eight out of 82 animals, a large number of labeled fibers passing through the lesion enclosed by the U-shaped wire were found. They were not collaterals sprouted from spared axons but evidently regenerated fibers because the BCX which is a complete crossing had been transected completely. By this procedure, the origin, course, and destination of the regenerated fibers were identified unambiguously.

Most of the regenerated fibers took a course similar to that of the normal projection and terminated in the normal projection areas, whereas a small proportion of them showed an aberrant course and termination. In the latter fibers the majority were uncrossed and terminated in the ipsilateral homologous structures of the normal projection.

Study of the time-course of regeneration revealed that the cut ends of axons began to swell as early as 15 min after transection, produced terminals tipped by growth cones in 14–24 hr, grew to cross the

251

lesion in 3 days, and distributed dense terminals in the thalamus by 19 days. The rate of axonal growth in successful regeneration was estimated to be 1–2 mm/day.

Functional connectivity of the regenerated cerebellothalamic projection was tested electrophysiologically in the same animals examined morphologically. Transection of the BCX abolished completely cerebellocerebral responses, which never reappeared in the animals in which axonal regeneration was abortive. However, in all animals in which marked axonal regeneration occurred, cerebellocerebral responses were evoked in the frontal motor and parietal association cortices as in intact animals. Thus, it is clear that the ability of growing axons to recognize the appropriate target and make functional connections during ontogenesis is well maintained during regeneration. Years of pessimism about the failure of mammalian central axons to regenerate are thus giving way to new optimism about their regrowth potential. Reconstruction of neural connections after damage to the brain may not be an impossible dream.

BRAIN SLICE PREPARATIONS IN NEUROSCIENCE
Comments on the Presentation by Professor Masanori Otsuka

CHOSABURO YAMAMOTO

Department of Physiology, Faculty of Medicine, Kanazawa University, Kanazawa 290, Japan

Professor Otsuka reported several marvelous preparations in which a tissue excised free from the mammalian central nervous system (CNS) maintains intact connections with peripheral organs and controls activities of the latter. Here, I compare features of Prof. Otsuka's preparations with those of brain slice preparations to clarify their merits and demerits.

The isolated nervous preparations of the CNS, either brain slices or the organized preparations referred to by Prof. Otsuka, have several advantages over the brain's *in vivo*. First, ionic composition of the extracellular solution can be easily changed in isolated preparations. Thus, the magnitude of synaptic transmission can be increased by increasing Ca^{2+} concentrations and reduced by increasing Mg^{2+} concentrations. The equilibrium potential of excitatory postsynaptic potentials can be altered by changes in concentrations of Na^+. Seizure discharges are induced by reductions in Cl^- concentrations. Second, chemical agents can be administered to the tissue at known concentrations. Thus, action potentials and synaptic potentials are suppressed by corresponding specific blocking agents. Third, movements caused by respiration and pulsation unavoidable in *in vivo* preparations are eliminated. This allows long, stable recording of electrical activities from single neurons.

Unfortunately, however, the isolated preparations of the mammalian CNS have a limitation. The size of these preparations is restricted to assure free diffusion of oxygen and glucose into, and of metabolites from, the innermost part of the tissues. Therefore, brain slices are usually cut to a thickness of 0.4 mm or less. Although this offers the additional advantage of making insertion of microelectrodes possible into a particular site or layer in the tissue under visual control, slices are too thin to include a complete functional circuit. Therefore, brain slices are suitable to study properties of single central neurons and of synaptic transmission between them but are not suitable to study how a neuronal circuit is organized to regulate a particular function.

For the study of functional circuits, the preparations reported by Prof. Otsuka are quite suitable. In order to overcome the limitation in size, he uses newborn rats or mice. With these preparations, therefore, it is difficult to study the functional circuits which develop later in life.

In summary, *in vitro* preparations allow various experiments which are impossible in *in vivo* brains. Brain slices are more appropriate to experiments at the single cell level, and the preparations of Prof. Otsuka will be useful to study the functional organization of brains of newborn animals.

SUBJECT INDEX

256

AUTHOR INDEX